Impact of biofilms in health

a transcriptomic perspective

Edited by

Nuno Cerca
University of Minho - Portugal

CENTRE OF BIOLOGICAL ENGINEERING

www.biofilmtranscriptomics.wordpress.com

Edições Universidade do Minho – Departamento de Engenharia Biológica

ISBN: 978-989-97478-6-9

1st Edition – January 2015
Braga, Portugal

Contents of this book

About this book

This book was written for undergraduate and graduate students in the field of microbial biofilms, basic molecular biology and transcriptomics, by presenting comprehensive information and detailed protocols optimized for Gram-positive and Gram-negative bacterial biofilm RNA extraction. The reader will learn how to work with complex microbial samples and how to quantify the genetic expression of genes of interest.

The book is divided in two main sections. The first addresses the principles behind gene expression studies, providing particular details on how to quantify gene expression from complex biofilm samples. The second section focuses on the impact of pathogenic biofilms in human health and reviews how transcriptomic studies help unveil some of the misteries behing biofilm pathogenicity.

This book has been adopted as the book course of an international practical workshop organized yearly by the Center of Biological Engineering, University of Minho, Braga, Portugal. More information can be found at:

www.biofilmtranscriptomics.wordpress.com

About the editor

Nuno Cerca is a Principal Investigator at the Centre of Biological Engineering of the University of Minho, Portugal (*www.ceb.uminho.pt*). He is an "Investigador FCT", a merit distinction awarded by the Portuguese Fundation of Science and Technology (*www.fct.pt*). During his research careear, he worked at Harvard Medical School, Boston, MA, USA and at Virginia Commonwealth University, Richmond, VA, USA. His current research interests are in the field of *Staphylococcus epidermidis* and *Gardnerella vaginalis* biofilms, with particular interest in gene expression and its relationship with bacterial virulence.

This book was co-funded by FCT Strategic Project PEst-OE/EQB/LA0023/2013, FCT project RECI/BBB-EBI/0179/2012 (FCOMP-01-0124-FEDER-027462) and by QREN, FEDER, ON2 project. NORTE-07-0124-FEDER-000027.

Introduction to Molecular Biology

Section I

Chapter 1

Introduction to bacterial molecular biology

Ana Henriques
CEB - Centre of Biological Engineering, LIBRO - Laboratory of Research in Biofilms Rosário Oliveira, University of Minho, Campus de Gualtar, 4710-057 Braga, Portugal
ana.henriques@deb.uminho.pt

Contents of chapter

Introduction to the chapter

The diversity seen in the many thousands of species present on Earth is without a doubt amazing, and many times overwhelming, even more so when the origin of this diversity is known. The source of this variability is encoded in the genetic information contained in each living cell that holds the blueprint for each organism.

The building blocks of molecular biology

The functional unit of genetic information is the gene, which in turn is composed of DNA (Deoxyribonucleic Acid) molecules. Although this is the primary source of genetic information it is not the only one, there are other molecules present in the cells that encode genetic information. As such, there are 3 main informational macromolecules in a cell: DNA, RNA (Ribonucleic Acid) and proteins (reviewed in Mattick 2003).

DNA is organized into genes which in turn are organized in chromosomes. The location and arrangement of the genetic information is one of the main differences between eukaryotic and prokaryotic cells. Eukaryotic cells (such as those of animals and plants) are characterized by the presence of membrane bound organelles within each cell, such as mitochondria, Golgi apparatus, endoplasmatic reticulum and the nucleus, and several chromosomes are present, localized in the nucleus of the cell, and are called the genome. In prokaryotes, like bacteria, the genome is comprised of a single chromosome which is found free on the cytoplam (Kerfeld, 2005).

Other genetic information units are plasmids, viral genomes (reviewed in Cohen 2013), organellar genomes and transposable elements (Buscher, 2012). Plasmids are DNA molecules which are independent of the bacterial genome and can replicate independently, they are not essential for the survival of the cell, however they often code for important traits such as antibiotic resistance (reviewed in Kroll 2010).

Viral genomes are another type of genetic information material. They can exist isolated, encapsulated in a viral capsid, or loose inside cells, they can be made up of DNA or RNA, which encodes the machinery that allows them to highjack the target cells as a way of replicating copies of the viruses (reviewed in Edwards 2005).

Organelar genomes can only be found in eukaryote cells as, prokaryotes do not possess intracellular organelles. Some examples of such genomes are those found in mitochondria, and plastids (genetic material found in plant chloroplasts) (Iborra, 2012).

Transposable elements are sequences of genetic material that can move themselves into new positions of the genome of a single cell. These movements can be of the 2 different types; "copy-paste" and "cut-paste", which can possibly lead to mutations and alteration in the total genome size (reviewed in Buscher 2012).

These units (plasmids, viral genomes, organellar genomes and transposable elements) are frequently used as tools in the genetic engineering of microorganisms specifically due to their non-essential nature that allows the construction of genetic sequences to be introduced into the genome that allow the introduction or deletion of genes (reviewed in Cohen 2013).

DNA Structure

Although DNA was identified for the first time in 1869, by Friedrich Miecher, its structure remained unknown for a further 80 years (reviewed in Dahm 2008). In the 1950s, several groups of researchers dedicated themselves to the study of this molecule. In 1953, Watson and Crick proposed the double helix model of DNA (Watson, 1953).

DNA is made up of 4 basic units, called nucleotides, Adenine (A), Guanine (G), Cytosine (C) and Thymine (T), which are organized in pairs, in the form of a double helix (Figure 1.1). The strands of DNA are antiparallel and have complementary nucleotide sequences. Each of the nucleotides pairs with only one another, A always pairs with T, and G always pairs with C (Brandenberg, 2011). This arrangement in pairs is very important for the determination of the physical and chemical properties of the DNA molecules.

Hydrogen bonds between DNA strands hold them together. The Adenine-Thymine pair (AT) has two hydrogen bonds whilst the Guanine - Cytosine (GC) pair has three hydrogen bonds. Because of this characteristic the GC bonds are stronger than those between AT, and thus require a larger amount of energy to break this bond. It is important to bear this in mind when considering *in vitro* applications involving nucleic acids, such as polymerase chain reaction (PCR) (Rychlik, 1990).

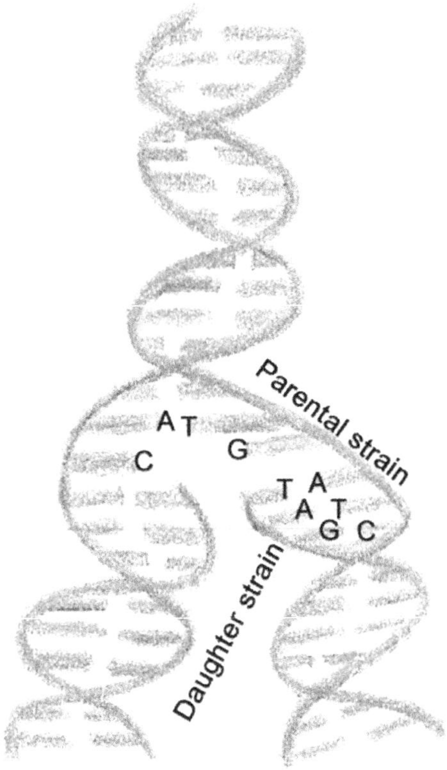

Figure 1.1 | DNA double helix. Note the complementary base pairing

The type of nucleotide pairing, allows the stacking of the base pairs which in turn increases the stability of the DNA molecule. As such DNA molecules with high GC, is more stable but not due to the presence of hydrogen bonds, but due to the stacking interactions between the base pairs;

this information is of particular importance when developing *in vitro* assays, (Yakovchuk, 2006).

A careful study of nucleic acid thermodynamics is beneficial for predicting the behavior of nucleic acid *in vitro*. The melting temperature (*Tm*) is defined as the temperature at which half of the DNA strands are in the random coil or single-stranded (ssDNA) state. In PCR this temperature is used as a guideline to calculate above which temperature the denaturation process occurs, and the optimal temperature for annealing of the primer to be used in the amplification (Breslauer, 1986).

The size of a DNA molecule is expressed in base pairs (e.g. the number of AT or GC pairs) where:

1,000 base pairs = 1 kilobase pairs = 1 kbp

1,000,000 base pairs = 1 megabase pairs = 1Mbp

In Nature, there is great range of genome sizes as can be seen in figure 1.2.

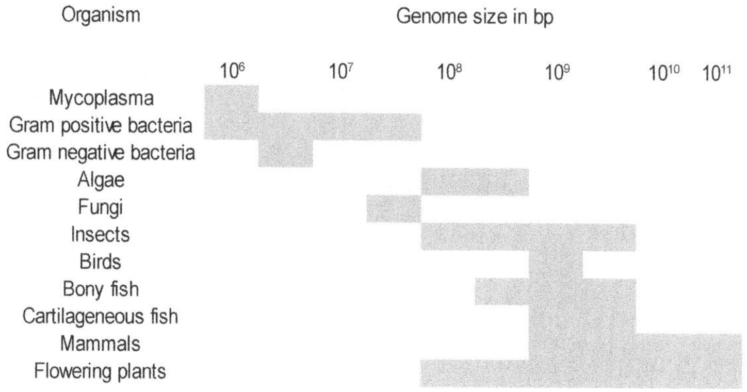

Figure 1.2 | Average genome size in living organisms

Besides the variability in the genome size between the different kingdoms (prokaryotes vs eukaryotes) it is possible to see that even within the same kingdom there is a wide variation of genome size as illustrated by the width of the bars in figure 1.2.

It should also be noted that not all genes are transcribed at the same time and in response to the same environmental stimuli and as such it is important to understand the flow of genetic information.

Genetic information flow

Genetic information flow can be divided into three stages:

1. **Replication**, where the DNA is duplicated,
2. **Transcription**: where information from DNA is transferred to RNA, that can be of 3 different types:
   - *mRNA* (messenger RNA): encodes polypeptides
   - *tRNA* (transfer RNA): plays role in protein synthesis
   - *rRNA* (ribosomal RNA): plays role in protein synthesis.
3. **Translation**: information in RNA is used to build polypeptides

The central dogma of molecular biology is the order of the genetic information flow: DNA is transcribed into RNA, which in turn is translated to proteins (reviewed in Li 2011). This information flow of genetic information in cells has been shown many times, and as such could be more appropriately be termed as Basic Principle of molecular biology (Brandenberg, 2011).

The information contained in the genetic material also needs to be maintained when the cells replicate and as such the replication of genetic information is also a very important basis for some molecular biology tools such as the polymerase chain reaction (PCR) (Saiki 1988).

As said previously, genetic information is stored in DNA that in turn is organized into genes, which make up the building blocks of the information contained in the cell. Genes can be identified by open reading frames, (ORFs), these are regions in the genome, where a start codon (see figure 1.6) signals the beginning of transcription and that finishes when a stop codon is found. Not all ORFs are necessarily transcribed, but its presence is often used as a tool for gene prediction (Shmatkov, 1999).

In order to signal the start of gene transcription promoter regions are often found. It contains the necessary sequences to activate or deactivate the gene transcription. Proximal promoter is considered to be 1000-50000 bp upstream the transcriptional start site (TSS) and are typically 100-1000 bp long. Promoters are also commonly used in biotechnology applications, for example in constructions of plasmids or inducible mutants (Estrem, 1999).

One promoter can induce transcription of several genes. The term operon is used to describe this set of genes controlled by the same promoter (reviewed in Jacob 2011).

Promoters can be inducible (respond to an environmental cue) or constitutive (allowing transcription in all occasions) (reviewed in Pátek 2013). In bacteria these sites are recognized by RNA polymerase. With inducible promoters, the transcription requires an activator molecular. One of the most studied and used promoter system if the Lac operon, where the activator molecule is lactose (figure 1.3) (Bohgdale, 2015).

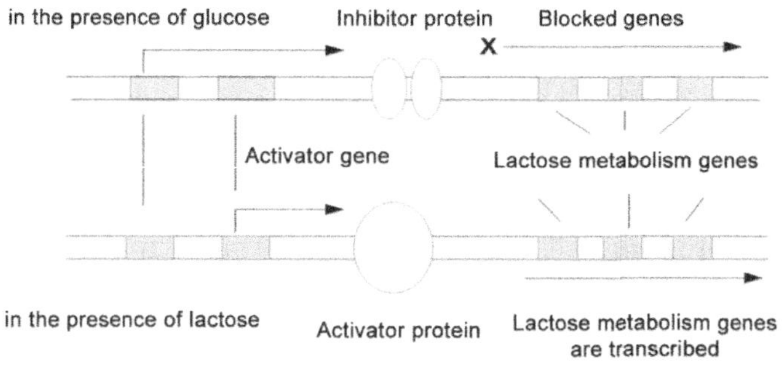

Figure 1.3 | Lac operon in bacteria. Briefly the lac operon is a two part control mechanism. In the absence of lactose (but in the presence of glucose), it activates the lac repressor gene that halts the production of the enzymes needed for the metabolism of lactose. In the presence of glucose. In the presence of lactose the activator gene is transcribed and this leads to the transcription of the lactose metabolism genes.

In prokaryotes, regulation of transcription is needed for the cell to quickly adapt to the ever-changing environment (Camejo, 2009). The presence or the quantity and type of nutrients determine which genes are expressed; in order to do that, genes must be regulated in some fashion. In prokaryotes, repressors bind to regions called operators that are generally located downstream from and near the promoter (normally part of the transcript). Activator or repressor molecules bind to the upstream portion of the promoter and determine whether a gene is transcribed in prokaryotes (Bohgdale, 2015).

Replication of genetic information

DNA replication is termed semi-conservative, which is that one of the double strains that make up the parental DNA double helix gives origin to a new DNA strand. A representation of the semiconservative replication of DNA can be seen in figure 1.1. The new DNA strand will be added to the initial parental strain, which works as a template for the addition of complementary precursors for each nucleotide called deoxynucleoside 5′ triphosphate (dNTP).

Each DNA strand has a polarity, or directionality which determines the orientation of nucleotides. The "front" end of the strand is called the *3' end* of the strand, and the "back" end is called the *5' end.* In double-stranded DNA, the paired strands are oriented in opposite directions: the 5'end of one strand aligned with the 3' end of the other. The polarity of a DNA strand is indicated by the direction of the arrow-like ribbon (Figure 1.4). The tail of the arrow represents the 5' end of the DNA strand, the head the 3' end (reviewed in Yelees 2014).

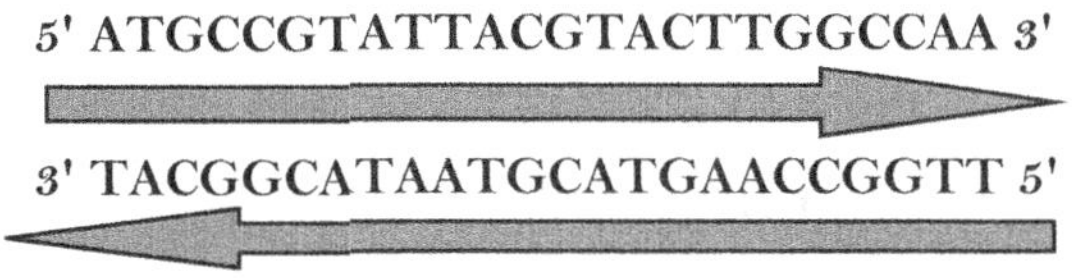

Figure 1.4 | Arrangement of DNA alignment. Information contained in DNA is always read from the 5' end to the 3' end.

Replication always proceeds from the 5′ end to the 3′ end, and the addition of the nucleotide precursors into the parental strain is catalyzed by the enzyme DNA polymerases, usually DNA polymerase II, and requires a RNA primer to initiate replication (Ivessa, 2013).

The Replication Fork

In prokaryotes DNA synthesis begins at the origin of replication, a sequence of nucleotides that signals the cell to initiate transcription of the information contained in the DNA into RNA.

In eukaryotes each gene is transcribed individually, whilst in prokaryotes clusters of genes can be transcribed together usually encoding for a set of proteins involved in the synthesis of a particular product. An mRNA encoding a group of cotranscribed genes is called a polycistronic mRNA or operon (reviewed in Wu 2014).

The replication of DNA begins at the replication fork, a zone of unwound DNA where replication occurs (reviewed in Kurt, 2013). The DNA unwinds with the action of topoisomerase and DNA helicase (number 7 and 6 respectively, in figure 1.5) that breaks the hydrogen bonds holding the two strand that become the leading and the lagging strands.

The **leading strand template** (number 2 in figure 1.5) is the strand of DNA being replicated continuously towards the replication fork. All DNA synthesis occurs 5'-3'. The original DNA strand must be read 3'-5' to produce a 5'-3' nascent strand.

The **lagging strand template** (number 1 in figure 1.5) grows in the direction opposite to the movement of the growing fork and it is synthesized discontinuously. Because the strand is growing away from the replication fork, it needs to be replicated in fragments because the DNA primase (that adds the RNA primer) has to wait until the fork opens to be able to put the primer (Figure 1.5) (reviewed in Yeeles 2014).

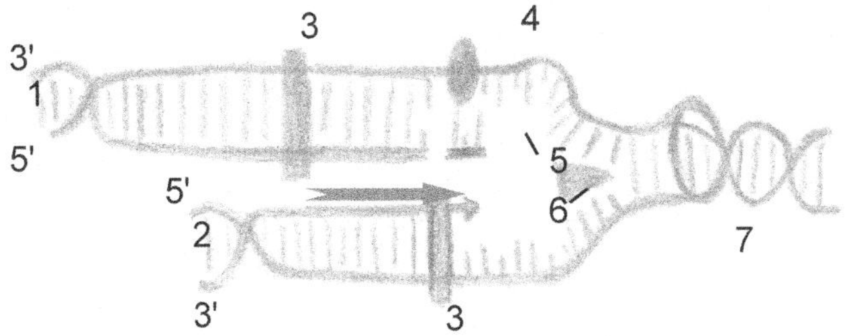

Figure 1.5 | Semi-conservative DNA Replication: (1) Lagging strand, (2) Leading strand, (3) DNA polymerase I, (4) DNA primase, (5) RNA primer, (6) Helicase, (7) Topoisomerase

RNA synthesis

Transcription is carried out by RNA polymerase, which uses DNA as a template and can be divided into 4 steps:

1. RNA polymerase unwinds the DNA strand
2. RNA nucleotide precursors: ATP, GTP, CTP, and UTP (instead of Thymine) are added to the chain
3. The Sugar backbone forms with the assistance of RNA polymerase

4. The hydrogen bonds between the DNA-RNA strand break, leaving the RNA strand free

Chain growth is 5′ to 3′ just like DNA replication but only one of the two strands of DNA is transcribed by RNA polymerase for any gene (or operon in some cases).

RNA polymerase recognizes DNA sites called promoters, DNA sequences that signal the sites for initiation of transcription which begins when both the RNA polymerase and a sigma factor bind to this region of the DNA. It should be noted that more than one gene can have the same promoter sequence and this can lead to their simultaneous transcription in response to the same environmental stimuli (Pátek, 2013).

The synthesis of RNA can be thought off as having 3 steps: initiation (already described in the paragraph above), elongation and termination. The elongation step is similar to that of DNA replication, as nucleotides are added by RNA polymerase to the open DNA chain in order to produce RNA molecules complementary to the DNA but where thymine is replaced by uracyl (Kingsford, 2007).

Each set of 3 nucleotides is called a codon and this set is translated into different proteins (figure 1.6). The possible combinations of nucleotides are the basis for genetic information required for the maintenance of cellular function (reviewed in Ermolaeva 2001).

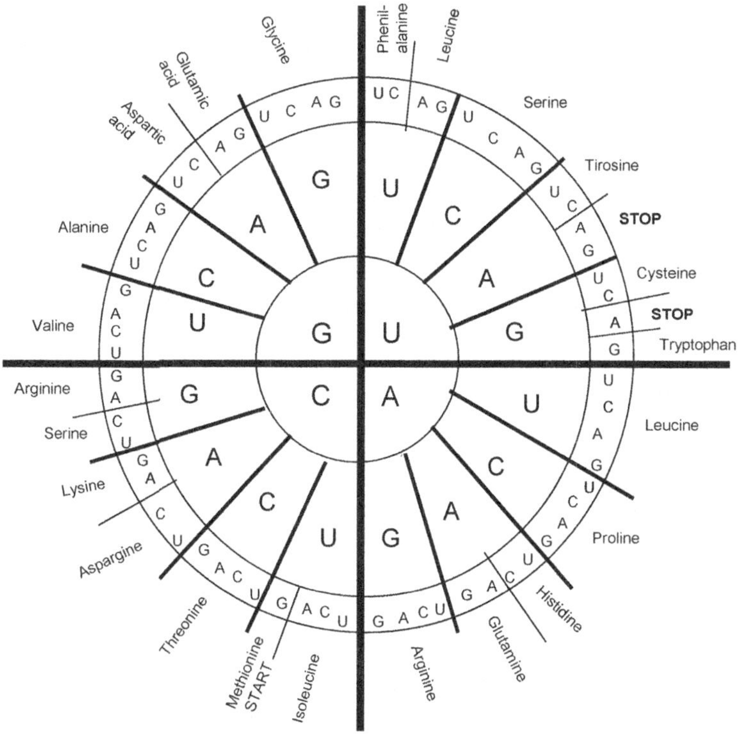

Figure 1.6 | The codon table. Each combination of 3 nucleotides is related to a specific aminoacid.

Transcription is terminated when the sigma factor is released which happens in prokaryotes when a "stop" codon is encountered (can be seen in figure 1.6) or when a transcription terminator is encountered (when the transcribed RNA forms a hairpin structure, see figure 1.7) (Kingsford, 2007). After transcription, the RNA has to be translated into proteins, and once again there is a strict table that dictates which RNA codons encode which proteins (figure 1.6), it can be observed that the same protein can be encoded by different codons (codon bias). Different factors have been proposed to be related to codon usage bias, including gene expression level, %G+C composition, amino acid conservation, RNA stability, optimal growth temperature among others (Heizer, 2006).

Figure 1.7 | RNA transcription physical terminators.

Most genes encode proteins, but some RNAs are not translated (i.e., rRNA, tRNA). Transfer RNA (tRNA) molecules help to deliver amino acids to the ribosome, where ribosomal RNA (rRNA) links amino acids together to form proteins. These last are very stable molecules whilst mRNA have short half-lives (only a few minutes) (reviewed in Dogini 2014). In prokaryotes there are 3 types of rRNA. In prokaryotes the 3 types of rRNA are: 23S, 16S and 5S (the S in the nomenclature stands the Svendberg unit for the sedimentation rate). The 16S rRNA of prokaryotes is highly conserved among bacterial species and is often used as a target for bacterial identification (reviewed in Dogini 2014).The understanding of the process through which genetic information in all living organisms is managed is the first step for a better understanding of the molecular biology tools that arose from the research done in this field.

Acknowledgments

AH holds a fellowship from the research project (NORTE-07-0124-FEDER-000027).

References

Bhogale PM, Sorg RA, *et al.* (2015) "What makes the lac-pathway switch: identifying the fluctuations that trigger phenotype switching in gene regulatory systems" in **Nucleic Acids Res.** 1;42(18):11321-8.

Breslauer, K.J.; Frank, R; *et al.* (1986). "Predicting DNA Duplex Stability from the Base Sequence" in **Proc. Natl. Acad. Sci. USA.** 83 (11): 3746–3750.

Bucher E, Reinders J, *et al.* (2012). "Epigenetic control of transposon transcription and mobility in *Arabidopsis*" in **Current Opinion in Plant Biology** 15 (5): 503–10.

Camejo A, Buchrieser C, *et al.* (2009) "*In vivo* transcriptional profiling of *Listeria monocytogenes* and mutagenesis identify new virulence factors involved in infection." in **PloS Pathogen** May;5(5):e1000449.

Cohen SN (2013) "DNA cloning: a personal view after 40 years" in **Proc Natl Acad Sci U S A.** 110(39):15521-9.

Dahm R (2008). "Discovering DNA: Friedrich Miescher and the early years of nucleic acid research" in **Hum. Genet.** 122 (6): 565–81.

Dogini DB, Pascoal VD, *et al.* (2014) "The new world of RNAs" in **Genet Mol Biol.** 37(1 Suppl):285-93.

Edwards RA and Rohwer F. (2005) "Viral metagenomics" in **Nature Reviews Microbiology** 3(6):504–10.

Ermolaeva MD (2001). "Synonymous codon usage in bacteria" in **Curr Issues Mol Biol.** 3 (4): 91–7.

Estrem, S. T.; Ross, W.; *et al.* (1999). "Bacterial promoter architecture: Subsite structure of UP elements and interactions with the carboxy-terminal domain of the RNA polymerase alpha subunit" in **Genes & development** 13 (16): 2134–2147.

Heizer EM Jr, Raiford DW, *et al.* (2006) "Amino acid cost and codon-usage biases in 6 prokaryotic genomes: a whole-genome analysis" in **Mol Biol Evol.** 23(9):1670-80.

Iborra FJ, Kimura H, *et al.* (2004). "The functional organization of mitochondrial genomes in human cells" in **BMC Biol**. 2: 9.

Ivessa AS.(2013) "Directionality of replication fork movement determined by two-dimensional native-native DNA agarose gel electrophoresis" in **Methods Mol Biol.**;1054:83-103.

Jacob, F. O. (2011). "The Birth of the Operon" in **Science** 332 (6031): 767–767.

Kerfeld, C. A.; Sawaya, M. R; *et al.*, (2005). "Protein structures forming the shell of primitive bacterial organelles" in **Science** 309 (5736): 936–8.

Kinsford CL, Ayanbule K, *et al.* (2007) "Rapid, accurate, computational discovery of Rho-independent transcription terminators illuminates their relationship to DNA uptake" in **Genome Biol**.;8(2):R22.

Kroll J, Klinter S, *et al.* (2010). "Plasmid addiction systems: perspectives and applications in biotechnology" in **Microb Biotechnol.** 3(6): 634–657.

Kurt, I. and O'Donell, M. (2013) "New insights into replisome fluidity during chromosome replication" in **Trends in Biochemical Science** 38(4): 195–203.

Li, G-W. and Xie, X. S. (2011) "Central dogma at the single-molecule level in living cells" in **Nature** 475(7356): 308–315.

Mattick JS. (2003) "Challenging the dogma: the hidden layer of non-protein-coding RNAs in complex organisms" in **Bioessays**. 25(10):930-9.

Pátek M, Holátko J *et al.* (2013) "Corynebacterium glutamicum promoters: a practical approach." in **Microb Biotechnol**; 6(2):103-17.

Rychlik, W.; Spencer, W. J. et al. (1990). "Optimization of the annealing temperature for DNA amplification *in vitro*" in **Nucleic Acids Res.** 18 (21): 6409–6412.

Saiki, RK; Gelfand DH; *et al.* (1988). "Primer-directed enzymatic amplification of DNA with a thermostable DNA polymerase" in **Science** 239 (4839): 487–91.

Shmatkov, A.M., Melikyan, A.A. *et al.* (1999) "Finding prokaryotic genes by the 'frame-by-frame' algorithm: targeting gene starts and overlapping genes" in **Bioinformatics**, 15, 874-886.

Watson J.D. and Crick F.H.C. (1953). "A Structure for Deoxyribose Nucleic Acid" in **Nature** 171 (4356): 737–738.

Wu, Z., Liu, J. *et al.* (2014) "DNA replication origins in archea" in **Frontiers in Microbiology**, volume 5, article 179.

Yakovchuk P, Protozanova E, *et al.* (2006). "Base-stacking and base-pairing contributions into thermal stability of the DNA double helix" in **Nucleic Acids Res.** 34 (2): 564–74.

Yeeles JT. (2014) "Discontinuous leading-strand synthesis: a stop-start story." in **Biochem Soc Trans.** 42(1):25-34.

Chapter 2

From single-gene to transcriptomic analysis

Ângela França
CEB - Centre of Biological Engineering, LIBRO - Laboratory of Research in Biofilms Rosário Oliveira, University of Minho, Campus de Gualtar, 4710-057 Braga, Portugal
afranca@deb.uminho.pt

Contents of chapter

Introduction to the chapter

In order to live, bacteria must respond quickly to the changes in the involving environment. Contrary to the genome, that remains almost unchanged over time, cells' transcriptome, the complete set of RNA molecules, is constantly changing as it responds to external stimuli. Hence, by studying the complete set of RNA molecules, it is possible to identify the genes that are being induced (needed), or repressed (not needed), in response to a particular condition. It is not surprising, therefore, that the study and quantification of RNA molecules has been considered as one of the most relevant tools in modern molecular biology. Indeed, it has been used in a myriad of applications, ranging from basic science (Best, 2013; Schwanhausser, 2011), through biotechnology (Raman, 2011; Rossouw, 2010), to clinical diagnosis (Goerke, 2001; Woods, 2013). In the context of bacterial biofilms, the study of RNA molecules has proved to be very useful, among others, in the study of biofilms phenotypic particularities (Carvalhais, 2014), tolerance to antibiotics (Williamson, 2012) and to explore the complex host-pathogen interaction (Franca, 2014), having provided very insightful information.

But how can researchers study and quantify gene expression? Nowadays, there are numerous methods that allow the study of gene expression. These methods are convened into two different groups: (i) targeted or (ii) genome-wide analyses. The use of the adequate approach will depend on the question that you intend to answer. When you already know the genes of interest and you need to analyze several samples, targeted methods, which allow the study of a small set of genes, are commonly used (Plexpress). On the other hand, when the genes of interest are not known, or when working with complex biological systems, an initial exploratory phase is often required Hence, genome-wide analyses, which allow the study of all, or high number of genes, are preferred (Plexpress). These techniques fall within one of the most stable "omic" sciences: transcriptomics. In figure 2.1 you can see the principal advantages and disadvantages of both targeted and genome-wide analysis.

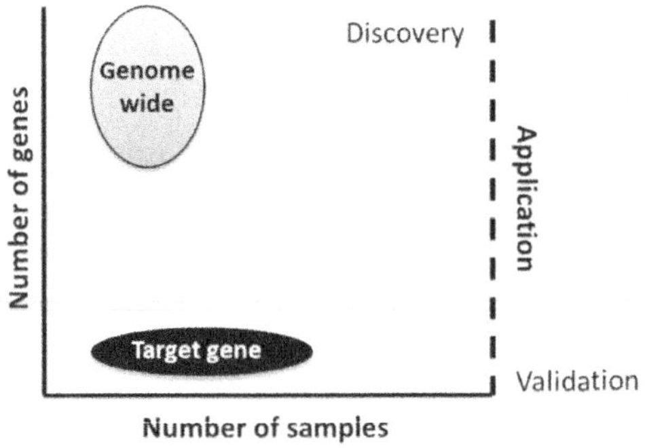

Figure 2.1 | Comparison between targeted and genome-wide analyses. This figure illustrates the major advantages and limitations of the two principal groups of gene expression quantification methods.

The progress of targeted and genome-wide analyses

For many years, only targeted methods were used for gene expression quantification studies. These techniques were, however, reductionists since in targeted analysis only a small number of genes could be analyzed (Fang and Casadevall, 2011). The study of complex systems was therefore achieved by reducing them to the study of individual molecules or pathways (Van Regenmortel, 2004). Although, groundbreaking discoveries were made using targeted analyses, it was largely hypothesis-driven (Mazzocchi, 2012). Hence, it was necessary to move forward toward a more integrative and complex analysis system.

In the late '90s, due to the advent of one of the first transcriptomic technologies, DNA microarrays, it became possible to perform, for the first time, large-scale studies (reviewed in Jaluria, 2007). This has allowed exploring, in a single analysis, the role, relationship and action of several transcripts that constitutes a tissue, organ or organism.

Nevertheless, concomitantly with the great development and uprising of transcriptomic technologies, targeted methods, due to their convenience, lower cost and their usage in genome-wide analysis validation, had also been improved over the years (Mazzocchi, 2012). Henceforward, targeted methods still hold remarkable influence in up-to-date molecular biology tools.

Within targeted and genome-wide analysis there are several molecular tools that can be used. In this chapter, a chronological perspective, as well as a brief description of the principles of the most widespread methods will be presented.

Targeted gene expression methods

Targeted gene expression methods are divided into two groups: (i) hybridization-based methods such as northern blot (Alwine, 1977) and (ii) PCR-based methods such as reverse transcriptase (Rappolee, 1988a) and quantitative PCR (Higuchi, 1993).

Northern blot

Until late '90s, mRNA measurements were routinely assessed using Northern blot, which was first introduced in 1977 by Alwine and his collaborators (Alwine, 1977). This method involves the isolation, fractioning and separation of RNA molecules in a denaturing gel electrophoresis. Then, the fragmented RNA molecules are transferred into a membrane followed by hybridization with a probe, a small DNA or RNA sequence specific for the gene of interest (Krumlauf, 1994). Based on the intensity of the bands, the abundance of target transcripts among samples is determined. Northern blot does not only detect the presence of the target or its abundance among different samples but also indicates size and RNA integrity. Indeed, is the only method that gives you such information (reviewed in Sharkey, 2004). However, despite its simplicity, due to RNA susceptibility to degradation during electrophoresis, the high quantity of RNA needed, and the difficulty of using multiple probes together (Streit, 2009), other alternative methods were developed in the meantime. Yet, some improvement have been made and northern blot is still used in today's molecular biology studies (Bhardwaj, 2012; Streit, 2009).

Reverse transcriptase-polymerase chain reaction

Polymerase chain reaction was conceived in 1983 by Kary Mullis (Mullis, 1986). Since then daily practices of basic sciences, medical diagnosis and even courts of law were utterly transformed. The revolutionizing character of PCR method is based on its capacity to synthesize millions of copies of DNA from virtually everywhere and from as much as a single cell (Mullis, 1990). This synthesis is accomplished by performing several heating and cooling cycles in the presence of key molecules and enzymes (Mullis, 1990).

During the PCR reaction there are three amplification phases (figure 2.2). In the beginning of the reaction there is an exponential amplification phase where, at each cycle, the amount of product is doubled (assuming 100% amplification efficiency). Subsequent, the linear phase takes place. Here, the amplification slows down since essential reagents are being consumed. This phase is characterized by high levels of variability between samples as the exhaustion of the reagents occurs at different rates in distinct samples (figure 2.2, note the differences in the linear and plateau phase between sample A and B). Finally, the reaction stops completely due to several limiting factors (Kainz, 2000) and no more products are synthesized. This last phase is termed plateau phase, or also called end-point phase, as the detection of the reaction products is traditionally performed at this point.

The PCR reaction is a three-step process. In the first cycle the temperature is usually raised to 95 °C in order create single-stranded DNA molecules. This process is called denaturation. Secondly, the temperature is decreased to 50-60ºC to allow the primers to bind to a specific region, of both sense and anti-sense strands, of the DNA target molecule. Primers are small sequences of DNA or RNA that trigger the synthesis of new DNA strands as they provide the 3' hydroxyl group (3'- OH) end needed for DNA polymerase to act. This process is called annealing.

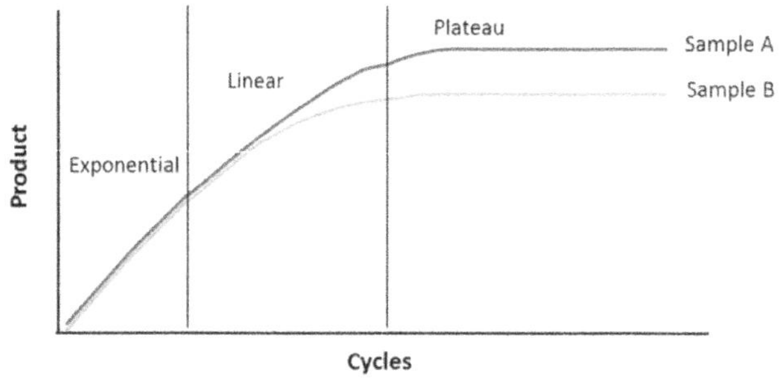

Figure 2.2 | PCR phases of amplification. As it can be seen, linear and plateau phases are highly variable among samples. Even among replicas of the same sample high variability may be observed.

Thereafter, DNA polymerase extends the primers by adding the nucleotides (adenine, thymine, guanine and cytosine), in a sequential manner, using the target DNA region as a template. This step is called extension or elongation and normally occurs at 72 ºC, the optimal temperature for DNA polymerase to work. The newly synthesized products serve as template in the next cycle creating, in this way, a chain reaction, giving rise to millions of copies of the target region. These denaturing, annealing and extension steps are then repeated, usually between 35 to 40 times. Finally, the products are loaded into an agarose gel where they are separated by their size.

For products visualization, the gel is stained with an intercalating dye, typically ethidium bromide that binds to double stranded molecules. At the end, using proper equipment, ethidium bromide can be excited and the emitted fluorescence enables the visualization and analysis of the results (Mullis, 1986; Mullis, 1990). Several modifications or variants were developed over the years in order to improve performance and specificity, and to attain the amplification of other molecules of interest such as RNA.

For RNA amplification and quantification, PCR was combined with the ability of the retroviral enzyme reverse transcriptase (RT) to convert mRNA molecules into complementary (c) DNA. This method is known as RT-PCR and has unquestionably revolutionized the way gene expression quantification assays were made since its introduction (Rappolee, 1988a; Rappolee, 1988b). This method involves two different stages: (i) the synthesis of cDNA in a reaction catalyzed by the reverse transcriptase and in the presence of primers, RNase inhibitors and nucleotides and, finally (ii) using PCR, cDNA molecules are exponentially amplified and quantified in an agarose gel.

Nevertheless, over the years, the quantitative capacity of RT-PCR has shown to be limited (Kainz, 2000). Assuming that PCR doubles the quantity of target molecule at each cycle, the relative differences in the initial quantity of target molecules can be determined based on the intensity of bands visualized on the agarose gel. This assumption however has a major limitation. Let's look back to the three amplification phases of the PCR reaction showed in figure 2.2. In the linear phase, the amount of product in the reaction is highly variable between samples, being no longer proportional to the amount of starting material (Kainz, 2000). The exhaustion of reagents occurs at different rates in different samples and hence, different samples start to diverge in their products quantities during the linear phase. Henceforth, relative amounts of amplified DNA can only be compared when the reactions are still in the exponential phase of amplification (Kainz, 2000). Nevertheless, if proper optimizations are performed to ensure that the reaction is stopped within the exponential phase, RT-PCR can then be safely used for gene expression quantification assays (Marone, 2001).

There are 2 main RT-PCR variations: the competitive and the relative RT-PCR. Competitive RT-PCR was developed for absolute quantification (Gause and Adamovicz, 1994). In order to do so, an external RNA molecule (the competitor) is synthesized and introduced into the sample. Importantly, the control molecule has to contain the same primer binding sequences as the target but has to give rise to a different size product. The control is then diluted into a standard curve of known concentration, and introduced into replicates of the sample. At the end, the amount of the transcript in the sample is determined comparing the intensity of the band of the target molecule to the one of the competitor, which concentration is known (Celi, 1993; Gause and Adamovicz, 1994).

In the case of relative qPCR, there are some particular requirements. This method involves the amplification of an endogenous control. This control allows the direct comparison of relative transcripts abundance as it provides samples normalization (Gause and Adamovicz, 1994). Nevertheless, the expression of the internal control chosen must be constant across all samples for the results to be accurate and meaningful. In addition, as referred before, comparison between samples in end-point analysis need to be performed in the exponential phase (Pfaffl, 1998). In this case, the exponential phase of amplification needs to be experimentally determined for both target and endogenous control. Finally, assessing both bands intensities, we can calculate the relative fold change of the target gene comparing to the endogenous control (Marone, 2001). However, endogenous controls are normally much more expressed than the gene of interest originating inaccurate comparisons. Hence, techniques with higher flexibility, allowing higher dynamic range of quantification were necessary.

Quantitative PCR

Since its introduction, in 1993, by Higuchi and his collaborators (Higuchi, 1993), quantitative (q) PCR, or also known as real time PCR, has been widely used. The principles of qPCR are very similar to the ones of RT-PCR. In both cases the synthesis of cDNA molecules is firstly performed and then cDNA molecules amplified by PCR.

The novelty of this technique is that combines both amplification and detection in one single step, since data is collected as the process occurs (Pfaffl, 2004). This real time collection of data is accomplished by the use of fluorescent chemistries that bind to the PCR products being synthesized (reviewed in Wong and Medrano, 2005). There are two levels of fluorescence intensities, which are both defined in the exponential phase of amplification (figure 2.3).

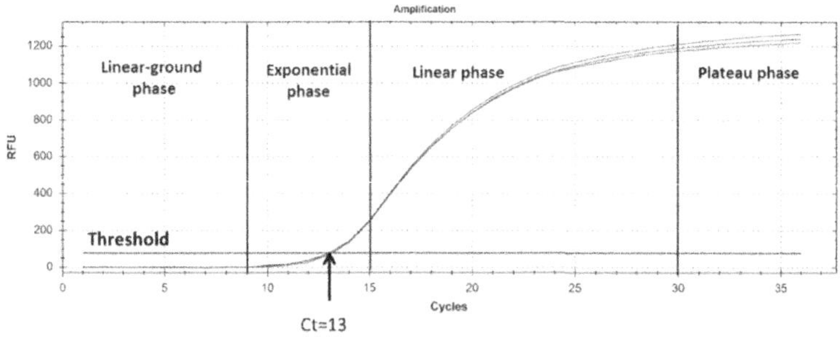

Figure 2.3 | qPCR phases and its associated concepts.

The baseline is the level where fluorescence has not arisen above the background signal. This is usually between the cycle 10 to 15 (reviewed in Wong and Medrano, 2005). When fluorescence is significantly higher than the background signal, the cycle threshold (Ct) can be defined (figure 2.3, indicated by the arrow). Ct is, thus, the PCR cycle when fluorescence intensity exceeds the background and is the key to calculate gene expression (Heid, 1996). As the fluorescent signal is inversely proportional to the amount of starting target molecules, the Ct will be lower in samples with higher amount of target molecules (Heid, 1996). As can be seen in figure 2.4, sample A has more initial amount of target molecules than samples B or C as the fluorescent has beaten the background fluorescence sooner. Nevertheless, Ct values *per se* are not meaningful. Several mathematical models have to be applied to accurately quantify gene expression (Livak and Schmittgen, 2001; Pfaffl, 2004).

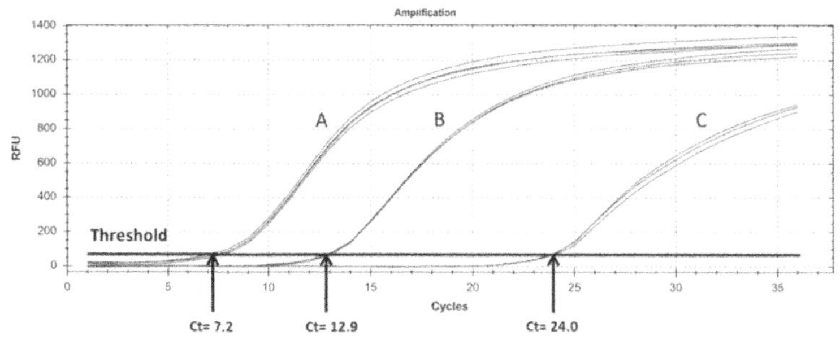

Figure 2.4 | Differences obtained in Ct values with samples with different in initial amount of target sequence. RFU-relative fluorescence units; Ct-cycle threshold.

Using qPCR you can determine both absolute number of targets (absolute quantification) or determined the relative quantify of a target molecule in different samples as explained, in detail, in chapter 4. qPCR presents several advantages including:

1. Higher sensitivity and reproducibility than all other targeted methods;
2. Large dynamic range of quantification enabling the detection from single molecules up to 10^9 molecules of the same target sequence in a single experiment;
3. Fast and easy to perform;
4. Allows the analysis of a larger numbers of samples;
5. No need for post-amplification handling, decreasing contamination issues and saving time (reviewed in Bustin, 2000 and Wong and Medrano, 2005; Schmittgen, 2001).

However, as any other technique, qPCR has disadvantages, such as the expenses associated with the equipment and reagents purchase. In addition, due to its high sensitivity, the presence of inhibitors such as salts and alcohol, or even pipetting errors may significantly impact the reliability and accuracy of final results. Nevertheless, if all precautions are taken, the advantages significantly overcome the limitations detected so far. Indeed qPCR has been considered the golden standard for confirmation and validation of the results obtained by genome-wide analysis (Carvalhais, 2014; Franca, 2014).

Before star a qPCR assay, please read MIQE Guidelines: Minimum Information for Publication of Quantitative Real-Time PCR Experiments. These guidelines were compiled in an attempt to ensure that qPCR assays are correctly performed throughout different laboratories and to bring clarity and transparency to qPCR based publications (reviewed in Bustin, 2009).

Genome-wide methods

Within genome-wide methods you may choose between 2 major groups: (i) hybridization- or (ii) sequence-based approaches (reviewed in Wang, 2009) being, respectively, DNA microarrays and RNA-sequencing the leading technologies of each group.

Microarrays

DNA microarray technology represents one of the greatest breakthroughs in genome-wide analyses (reviewed in Jaluria, 2007). It opened new precedents in molecular biology research, allowing the determination, on a single analysis, of the expression levels of thousands of genes (Morey, 2006). Although the principles behind microarrays technology were conceived short after Southern blot, in 1975 (Southern, 1975), it only gained widespread attention when researchers were able to obtain reproducible results (Brown and Botstein, 1999; Li, 2002; Nakanishi, 2001). The first paper using DNA microarrays analysis was published by Schena and colleagues in 1995 (Schena, 1995).

DNA microarrays are assemblies of microscopic probes, DNA sequences corresponding to a particular gene, immobilized onto a solid surface or membrane, where target sequences will then bind specifically by a process called hybridization (reviewed in Li, 2002). Target sequences are labeled with specific dyes for identification and for later quantification. The quantification of target molecules are determined by measuring the signal emitted by the labeling dye bound (reviewed in Butte, 2002 and Li, 2002 and Nakanishi, 2001). This procedure is challenging and includes several steps such as:

1. Microarrays slides manufacturing. Opportunely, nowadays, several slides for the most popular microorganism are already ready-to-use;
2. Total RNA extraction;
3. Genomic DNA removal;
4. Incorporation of fluorescent dyes. Usually Cy3 and Cy5 are the dyes used;
5. Hybridization, incubation and washing;
6. Generation of arrays-images using fluoresce microarrays proper scanners;

7. Extracting signal intensities from images. Fluorescent intensities, of each single spot, are measured and recorded alongside with the spot coordinates and specific transcript identifier. The intensity of the generated signal depends on the amount of targets molecules bound to the probe-molecules within one spot;
8. Data normalization and data analysis. Dealing with that high amount of data requires established bioinformatics tools for normalization, statistical analyses and gene ontology and pathway analysis (reviewed in Jaluria, 2007; Leung and Cavalieri, 2003).

Before starting a microarray experiment, it would be important to read the MIAME guidelines: Minimum Information About a Microarray Experiment. This document outlines the information need to perform a microarray experiment as well as the information that should be provided when describing it. These guidelines will ensure that microarrays experiments were correctly performed and that can be unambiguously reproduced(Brazma, 2001).

Despite its tremendous success, microarray technology, as any of the other techniques present their own limitations. Background hybridization is one of the most important, as it influences the accuracy of transcripts measurements, in particular, transcripts with low abundances (Zhao, 2014). Moreover, probes are limited to known genes, making it difficult the study of novel sequences (Marioni, 2008). To overcome these limitations, sequencing-based technologies were introduced for gene expression quantification assays.

RNA-sequencing

In the late '70, working separately, Frederick Sanger (Sanger, 1977) and Walter Gilbert (together with graduate student Allan Maxam) (Maxam and Gilbert, 1977) developed two methods for DNA sequencing. Due to its simplicity and efficiency, Sanger method was more successfully and was widely used around the world (Shendure and Ji, 2008). However, limitations associated with throughput, scalability, speed and resolution have triggered the development of new technologies (reviewed in Shendure, 2004).

Hence, sequenced-based methods have undergone many changes and improvements over the years, giving rise, in 2005, to next-generating sequencing technologies (Margulies, 2005; Shendure, 2005). The principle is the same: the sequences of small fragments of DNA are identified from the signals emitted when nucleotides are incorporated during the re-synthesis of the DNA template. However, next-generation sequencing, rather than being limited to a single or few DNA fragments, extends this process across millions of reactions, in a massively parallel fashion, enabling the sequencing of entire genomes, transcriptomes or epigenomes (reviewed in Liu, 2012).

RNA-sequencing is one of the most successfully next-generation sequencing technologies used for the study of transcriptomes (reviewed in Wang, 2009). The power of RNA-sequencing in quantifying cells' transcriptome is remarkable. By obtaining several millions of short sequence reads from the transcripts of a condition of interest, RNA-sequencing is able to detect and quantify RNA, including those with low and high abundance (Mortazavi, 2008). As the capacity of RNA-sequencing to generate several hundred million reads continues to increase, it will be possible to analyzed much more complex samples (Zhao, 2014).

RNA-sequencing name can create some misperception as indeed, it does not sequence mRNA itself but its cDNA copy. However, direct RNA sequence are being already developed in order to overcome some bias involved with cDNA synthesis (reviewed in Ozsolak and Milos, 2011). This methodology is very laborious and involves several steps including:

1. Total RNA isolation;
2. Genomic DNA removal;
3. Depletion of rRNA. In the cases of host-pathogen interaction, host RNAs need to be specifically removed;
4. Preparation of cDNA libraries;
5. Amplification of cDNA libraries;
6. Sequencing of cDNA libraries. Short sequences, typically between 30–400 bp, from one end (single-end sequencing) or both ends (pair-end sequencing) are sequenced;
7. Data analysis. The resulting reads are aligned to a reference genome or reference transcripts and then normalized to correct sequence error, nucleotide composition, and length or library preparation biases (reviewed in Wang, 2009).

The quantification of gene expression is then accomplished by correlating the amount of starting material with the frequency of each sequence read versus the number of total reads obtained. In addition, other normalizations may also be required, in order to account for differences in gene or sequence read lengths (Mortazavi, 2008). The best method for gene expression quantification by RNA-sequencing is, however, still a matter of discussion, as constant advancements are being made (Lee, 2011).

Nevertheless, using RNA-sequencing we are able to overcome some of the most pertinent limitations of hybridization-based methods such as microarrays. RNA-sequencing is more flexible as it is not limited to detecting transcripts that correspond to existing genomic sequences (Zhao, 2014). In this way, RNA-sequencing enables the detection of completely novel transcripts and hence, it is possible to study microorganism with unsequenced genomes (Zhao, 2014). In addition, it offers increased specificity and sensitivity. Sequencing coverage depth, or the percentage of transcripts measured, can be easily increased to detect weakly expressed genes (reviewed in Wang, 2009). Moreover, RNA-sequencing can accurately identify and quantify both rare and common transcripts, within more than six orders of magnitude of dynamic range (Zhao, 2014).

Yet, despite its advantages, the costs associated with RNA-sequencing are still high and are currently limiting its engagement, particularly for large scale studies. However, while the costs of microarray assay are almost fixed, the ones associated with RNA-sequencing are declining meanwhile the number of reads per experiment is increasing. Hence, in the future it is expected that RNA-sequencing technology can be affordable for most researchers and become the tool of choice for transcriptomic analysis. Some interesting approaches such as the multiplexing of different libraries, by using small nucleotide sequences called barcodes to tag each library (Smith, 2010), have significantly decreased the costs associated with RNA-sequencing analysis. However, in this case, several factors such as the size of the genome of the microorganism, pretended coverage, the number of reads given by the equipment used and if single-end or pair-end reads will be performed, need to considered before to determine the number of libraries that can be multiplexed (Zhao, 2014).

Since RNA-sequencing is still in its infancy when compared with qPCR or microarrays, contrary to what is observed for these methods, no guidelines for RNA-sequencing experimental design and data analysis were created so far. However there are several good review papers that can guide you throughout your experiments (Haas and Zody, 2010; Mortazavi, 2008; Wang, 2011; reviewed in Wang, 2009).

Conclusions

As you have learned throughout this chapter, there are several molecular tools available for gene expression quantification. However, care should be taken when selecting the appropriated method for your assays.

Nevertheless, the most advantageous techniques, within targeted and genome-wide analyses, are, without any doubt, qPCR and RNA-sequencing methodologies.

Acknowledgments

AF holds a fellowship from the research project EXPL/BIA-MIC/0101/2013 (FCOMP-01-0124-FEDER-041246).

References

Alwine JC, Kemp DJ *et al.* (1977) "Method for detection of specific RNAs in agarose gels by transfer to diazobenzyloxymethyl-paper and hybridization with DNA probes" in **Proc Natl Acad Sci U S A**, 74:5350-5354

Best JA, Blair DA *et al.* (2013) "Transcriptional insights into the CD8(+) T cell response to infection and memory T cell formation" in **Nat Immunol**, 14:404-412

Bhardwaj AR, Pandey R *et al.* (2012) "Northern blot analysis for expression profiling of mRNAs and small RNAs" in **Methods Mol Biol**, 883:19-45

Brazma A, Hingamp P *et al.* (2001) "Minimum information about a microarray experiment (MIAME)-toward standards for microarray data" in **Nat Genet**, 29:365-371

Brown PO and Botstein D (1999) "Exploring the new world of the genome with DNA microarrays" in **Nat Genet**, 21:33-37

Bustin SA (2000) "Absolute quantification of mRNA using real-time reverse transcription polymerase chain reaction assays" in **J Mol Endocrinol**, 25:169-193

Bustin SA, Benes V *et al.* (2009) "The MIQE guidelines: minimum information for publication of quantitative real-time PCR experiments" in **Clin Chem**, 55:611-622

Butte A (2002) "The use and analysis of microarray data" in **Nat Rev Drug Discov**, 1:951-960

Carvalhais V, Franca A *et al.* (2014) "Dormancy within *Staphylococcus epidermidis* biofilms: a transcriptomic analysis by RNA-seq" in **Appl Microbiol Biotechnol**, 98:2585-2596

Celi FS, Zenilman ME *et al.* (1993) "A rapid and versatile method to synthesize internal standards for competitive PCR" in **Nucleic Acids Res**, 21:1047

Fang FC and Casadevall A (2011) "Reductionistic and holistic science" in **Infect Immun**, 79:1401-1404

Franca A, Carvalhais V *et al.* (2014) "Alterations in the *Staphylococcus epidermidis* biofilm transcriptome following interaction with whole human blood" in **Pathog Dis**, 70:444-448

Gause WC and Adamovicz J (1994) "The use of the PCR to quantitate gene expression" in **PCR Methods Appl**, 3:S123-S135

Goerke C, Bayer MG *et al.* (2001) "Quantification of bacterial transcripts during infection using competitive reverse transcription-PCR (RT-PCR) and LightCycler RT-PCR" in **Clin Diagn Lab Immunol**, 8:279-282

Haas BJ and Zody MC (2010) "Advancing RNA-Seq analysis" in **Nat Biotechnol**, 28:421-423

Higuchi R, Fockler C *et al.* (1993) "Kinetic PCR analysis: real-time monitoring of DNA amplification reactions" in **Biotechnology (N Y)**, 11:1026-1030

Jaluria P, Konstantopoulos K *et al.* (2007) "A perspective on microarrays: current applications, pitfalls, and potential uses" in **Microb Cell Fact**, 6:4

Kainz P (2000) "The PCR plateau phase - towards an understanding of its limitations" in **Biochim Biophys Acta**, 1494:23-27

Krumlauf R (1994) "Analysis of gene expression by northern blot" in **Mol Biotechnol**, 2:227-242

Lee S, Seo CH *et al.* (2011) "Accurate quantification of transcriptome from RNA-Seq data by effective length normalization" in **Nucleic Acids Res**, 39:e9

Leung YF and Cavalieri D (2003) "Fundamentals of cDNA microarray data analysis" in **Trends Genet**, 19:649-659

Li X, Gu W *et al.* (2002) "DNA microarrays: their use and misuse" in **Microcirculation**, 9:13-22

Liu L, Li Y *et al.* (2012) "Comparison of next-generation sequencing systems" in **J Biomed Biotechnol**, 2012:251364

Livak KJ and Schmittgen TD (2001) "Analysis of relative gene expression data using real-time quantitative PCR and the 2(-Delta Delta C(T)) Method" in **Methods**, 25:402-408

Margulies M, Egholm M *et al.* (2005) "Genome sequencing in microfabricated high-density picolitre reactors" in **Nature**, 437:376-380

Marioni JC, Mason CE *et al.* (2008) "RNA-seq: an assessment of technical reproducibility and comparison with gene expression arrays" in **Genome Res**, 18:1509-1517

Marone M, Mozzetti S *et al.* (2001) "Semiquantitative RT-PCR analysis to assess the expression levels of multiple transcripts from the same sample" in **Biol Proced Online**, 3:19-25

Maxam AM and Gilbert W (1977) "A new method for sequencing DNA" in **Proc Natl Acad Sci U S A**, 74:560-564

Mazzocchi F (2012) "Complexity and the reductionism-holism debate in systems biology" in **Wiley Interdiscip Rev Syst Biol Med**, 4:413-427

Morey JS, Ryan JC *et al.* (2006) "Microarray validation: factors influencing correlation between oligonucleotide microarrays and real-time PCR" in **Biol Proced Online**, 8:175-193

Mortazavi A, Williams BA *et al.* (2008) "Mapping and quantifying mammalian transcriptomes by RNA-Seq" in **Nat Methods**, 5:621-628

Mullis K, Faloona F *et al.* (1986) "Specific enzymatic amplification of DNA in vitro: the polymerase chain reaction" in **Cold Spring Harb Symp Quant Biol**, 51 Pt 1:263-273

Mullis KB (1990) "Target amplification for DNA analysis by the polymerase chain reaction" in **Ann Biol Clin (Paris)**, 48:579-582

Nakanishi T, Oka T *et al.* (2001) "Recent advances in DNA microarrays" in **Acta Med Okayama**, 55:319-328

Ozsolak F and Milos PM (2011) "RNA sequencing: advances, challenges and opportunities" in **Nat Rev Genet**, 12:87-98

Pfaffl M, Meyer HH *et al.* (1998) "Quantification of insulin-like growth factor-1 (IGF-1) mRNA: development and validation of an internally standardised competitive reverse transcription-polymerase chain reaction" in **Exp Clin Endocrinol Diabetes**, 106:506-513

Pfaffl MW (2004) "Quantification strategies in real-time PCR" p. 87-112. In SA Bustin (Eds.), A-Z of quantitative PCR, International University Line, La Jolla, CA, USA

Plexpress "Gene expression analysis: A review". http://www.plexpress.fi/tiedostot/Plexpress_Whitepaper.pdf (last accessed October 13th 2014)

Raman B, McKeown CK *et al.* (2011) "Transcriptomic analysis of *Clostridium thermocellum* ATCC 27405 cellulose fermentation" in **BMC Microbiol**, 11:134

Rappolee DA, Brenner CA *et al.* (1988a) "Developmental expression of PDGF, TGF-alpha, and TGF-beta genes in preimplantation mouse embryos" in **Science**, 241:1823-1825

Rappolee DA, Mark D *et al.* (1988b) "Wound macrophages express TGF-alpha and other growth factors in vivo: analysis by mRNA phenotyping" in **Science**, 241:708-712

Rossouw D, van den Dool AH *et al.* (2010) "Comparative transcriptomic and proteomic profiling of industrial wine yeast strains" in **Appl Environ Microbiol**, 76:3911-3923

Sanger F, Air GM *et al.* (1977) "Nucleotide sequence of bacteriophage phi X174 DNA" in **Nature**, 265:687-695

Schena M, Shalon D *et al.* (1995) "Quantitative monitoring of gene expression patterns with a complementary DNA microarray" in **Science**, 270:467-470

Schmittgen TD (2001) "Real-time quantitative PCR" in **Methods**, 25:383-385

Schwanhausser B, Busse D *et al.* (2011) "Global quantification of mammalian gene expression control" in **Nature**, 473:337-342

Sharkey FH, Banat IM *et al.* (2004) "Detection and quantification of gene expression in environmental bacteriology" in **Appl Environ Microbiol**, 70:3795-3806

Shendure J and Ji H (2008) "Next-generation DNA sequencing" in **Nat Biotechnol**, 26:1135-1145

Shendure J, Mitra RD *et al.* (2004) "Advanced sequencing technologies: methods and goals" in **Nat Rev Genet**, 5:335-344

Shendure J, Porreca GJ *et al.* (2005) "Accurate multiplex polony sequencing of an evolved bacterial genome" in **Science**, 309:1728-1732

Smith AM, Heisler LE *et al.* (2010) "Highly-multiplexed barcode sequencing: an efficient method for parallel analysis of pooled samples" in **Nucleic Acids Res**, 38:e142

Southern EM (1975) "Detection of specific sequences among DNA fragments separated by gel electrophoresis" in **J Mol Biol**, 98:503-517

Streit S, Michalski CW *et al.* (2009) "Northern blot analysis for detection and quantification of RNA in pancreatic cancer cells and tissues" in **Nat Protoc**, 4:37-43

Van Regenmortel MH (2004) "Reductionism and complexity in molecular biology. Scientists now have the tools to unravel biological and overcome the limitations of reductionism" in **EMBO Rep**, 5:1016-1020

Wang L, Si Y *et al.* (2011) "A low-cost library construction protocol and data analysis pipeline for Illumina-based strand-specific multiplex RNA-seq" in **PLoS One**, 6:e26426

Wang Z, Gerstein M *et al.* (2009) "RNA-Seq: a revolutionary tool for transcriptomics" in **Nat Rev Genet**, 10:57-63

Williamson KS, Richards LA *et al.* (2012) "Heterogeneity in *Pseudomonas aeruginosa* biofilms includes expression of ribosome hibernation factors in the antibiotic-tolerant subpopulation and hypoxia-induced stress response in the metabolically active population" in **J Bacteriol**, 194:2062-2073

Wong ML and Medrano JF (2005) "Real-time PCR for mRNA quantitation" in **Biotechniques**, 39:75-85

Woods CW, McClain MT *et al.* (2013) "A host transcriptional signature for presymptomatic detection of infection in humans exposed to influenza H1N1 or H3N2" in **PLoS One**, 8:e52198

Zhao S, Fung-Leung WP *et al.* (2014) "Comparison of RNA-Seq and microarray in transcriptome profiling of activated T cells" in **PLoS One**, 9:e78644

Chapter 3

Challenges of gene expression studies in biofilms

Nuno Cerca
CEB - Centre of Biological Engineering, LIBRO - Laboratory of Research in Biofilms Rosário Oliveira, University of Minho, Campus de Gualtar, 4710-057 Braga, Portugal
nunocerca@ceb.uminho.pt

Contents of chapter

Introduction to the chapter

For those researchers with experience in RNA extraction and gene expression quantification in planktonic bacterial cultures, it might came as a surprise the realization that biofilm cultures present some extra challenges and often, previously optimized protocols will not be efficient and/or reliable when applied in biofilm studies. This chapter will take you through a 4 year journey in the optimization of protocols to be applied in gene expression studies in biofilms, with particular relevance for gram positive *Staphylococcus epidermidis.*

Choosing the right RNA extraction kit

RNA extractions kits need to perform two essential functions: to lyse the bacteria and to separate the RNA. When it concerns the lysing step, gram positive bacteria will present a bigger challenge, due to the thickness and composition of the outer wall. In theory, mechanical and chemical lysis kits will, therefore, present and intrinsic advantage, over enzymatic kits. Regarding the separation of RNA, there are two main strategies: separation by affinity columns or by chemical separation; normally, the latest takes a longer period of time to perform (Franca, 2012b). Currently there are several commercial kits to choose from, including one dedicated to extract RNA from biofilms (http://www.mobio.com/biofilm-rna-isolation/). However, during the time most of the experiments described in this chapter were performed, the biofilm-specific kit was not commercially available.

As I wrote in the introduction of this chapter, I had previous experience in performing gene expression studies in planktonic cultures of both gram positive *Staphyloccous aureus (Cerca, 2008)* and gram negative *Escherichia coli (Cerca, 2008b)* and, therefore, I was surprised to see that in *S. epidermidis biofilms,* I wasn't able to extract high quality RNA. As I was curious to see if this was a *S. epidermidis* or a biofilm related problem, I've decided to analyze the yields of 5 different RNA extraction kits in 2 gram positive and 2 gram negative bacteria. The selection of bacteria was based on the research interests of members of the Biofilm Group at the University of Minho, at that time: *Salmonella enterica, Escherichia coli* and *Lysteria monocitogenes* (figure 3.1).

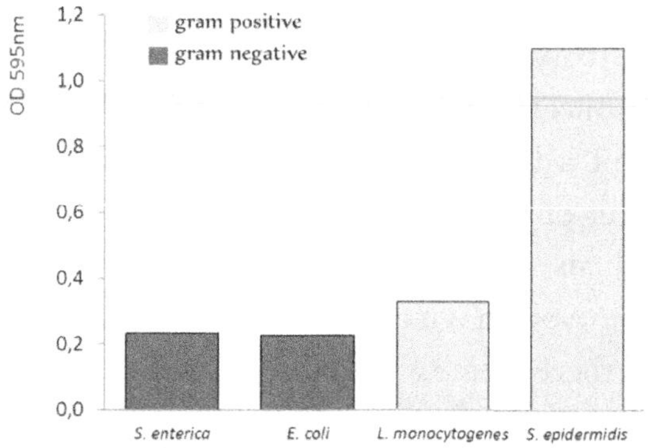

Figure 3.1 | Biofilm quantification by 4 distinct bacterial species (Franca, 2011; Franca, 2012a). Biofilms were formed in culture plates and quantified with crystal violet staining.

The kits tested had the following characteristics:

- Fast RNA Pro Blue (MP Biomedicals): chemical and mechanical lysis, chemical extraction;
- PureZol (BioRad): chemical lysis and extraction;

- RNAspin (GE Healthcare): enzymatic lysis and silica membrane affinity extraction;
- PureLink (Invitrogen): enzymatic lysis and silica membrane affinity extraction;
- GenJet (Fermentas): enzymatic lysis and silica membrane affinity extraction.

Interestingly, of all the commercial kits tested, few yielded high quality and high concentration RNA (figure 3.2).

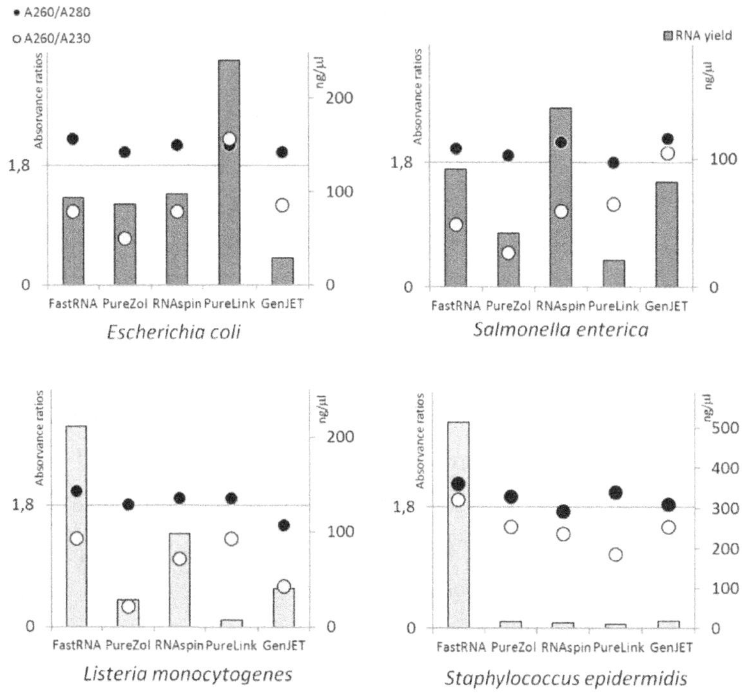

Figure 3.2 | Different commercial RNA extraction kits yield RNA with different quantity and quality (Franca, 2011; Franca, 2012a). For high quality RNA, the minimum desired concentration should be 40 ng/µl and the 260/280 and 270/230 absorbance ratios should be above 1.8.

Curiously, while from gram positive bacteria Fast RNA Pro Blue was clearly the best kit, gram negative bacteria performed better under other conditions: PureLink was the best kit for *E. coli* biofilms while RNA spin was the best kit for *S. enterica*.

Our next step was to quantify the expression on known genes, as compared with the reference gene 16S RNA and try to relate those results with the characteristics described above for the RNA samples. For *S. enterica* we selected the gene *prfA*, involved in the activation of several virulence genes (Rodrigues, 2011). For *E. coli*, we selected the gene *tyrB*, involved in amino-acid biosynthesis (Gelfand and Rudo, 1977). For *L. monocytogenes* the selected gene was *rpoS*, known for its role in the activation of the stress response (Rodrigues, 2011). Finally, *icaA* was the gene selected for *S. epidermidis*, for its dual role in biofilm accumulation (Heilmann, 1996) and evasion to the host immune system (Cerca, 2006).

As shown in figure 3.3, gene expression was influenced by the RNA used, independently of the bacterial strain tested. Interestingly, the differences observed were lower for the gram negative bacteria than from the gram positive bacteria, being *L. monocytogenes* the species with a higher variability in gene expression quantification. At this point, an important question arose: what would be the best kit for each biofilm model? We still need to perform many more experiments, before that answer could be giving. However, due to time and financial issues, we had to focus our research using one model and we chose *S. epidermidis*, the biggest biofilm forming species of the group tested (figure 3.1).

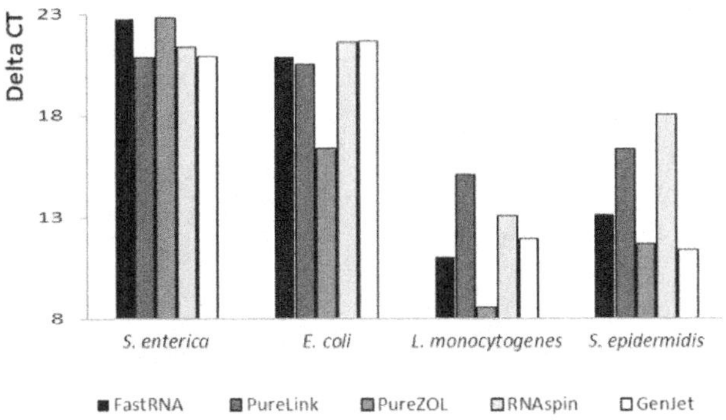

Figure 3.3 | The effect on gene expression using RNA extracted in the same experimental conditions, with different commercial RNA extraction kits (Franca, 2011; Franca, 2012a). For a stable gene expression assays, the difference between the cycle threshold of the reference gene and the target gene (Delta CT) should be constant.

Taking in consideration that biofilm formation varies not only between species but also within the same species (Cerca, 2005), our next experiments focused on the characterization of the biofilm and biofilm matrix formed by two different *S. epidermidis* clinical isolates. As shown in figure 3.4, *S. epidermidis* strain 1457 had almost the double of the total biomass of strain M187, and had four times more proteins in the matrix and about two times more polysaccharides. We then repeated the same experiments as before, and determined the RNA yield and quality from biofilms of the two different strains, using three RNA extraction kits with different characteristics (as described above). Not surprisingly, RNA Pro Blue was the best kit for both *S. epidermidis* strains. Interestingly, a trend was observed in the two non-mechanical lysis kits: it has harder to obtain RNA from the biofilms with higher biomass and higher protein and polysaccharide content in the matrix (figure 3.5).

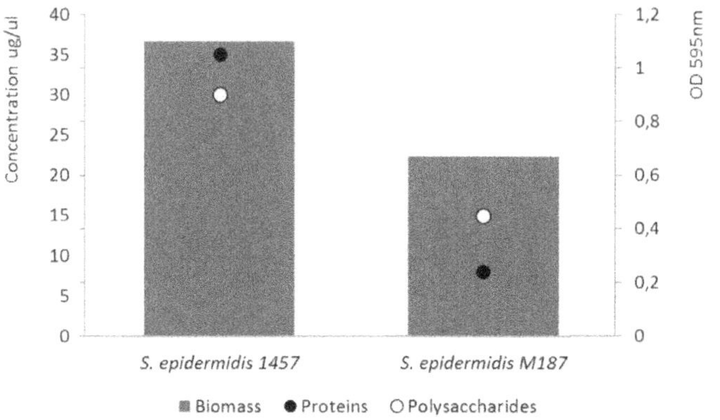

Figure 3.4 | Characterization of the biofilm and biofilm matrix of two distinct clinical isolates of *S. epidermidis* (Franca, 2011). Total biomass was determined by crystal violet staining; matrix proteins were quantified by the BCA protein assay and matrix polysaccharides were quantified by the Dubois method.

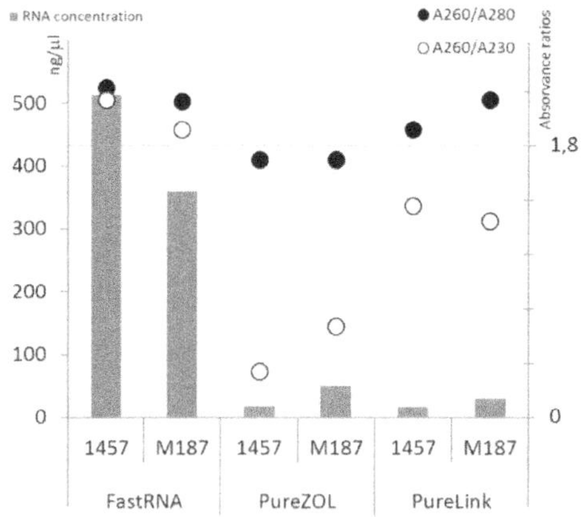

Figure 3.5 | Different commercial RNA extraction kits yield RNA with different quantity and quality (Franca, 2011). For high quality RNA, the minimum desired concentration should be 40 ng/µl and the 260/280 and 270/230 absorbance ratios should be above 1.8

As it became obvious to us that the extra-cellular components of the biofilm were, in fact, impairing RNA extraction, we performed one last experiment, aimed to determine the variability of gene expression in our two strains. However, we did not saw a strain-dependent effect but, as before, we detected significant variations on the gene expression normalization, having the PureLink kit an approximately ten-fold reduction in gene expression quantification, as compared with the other kits (figure 3.6).

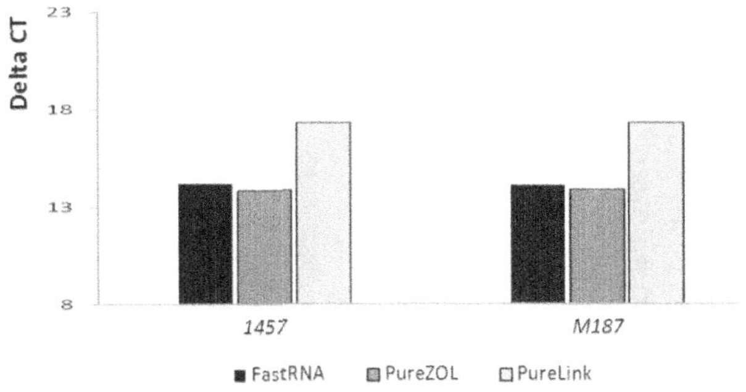

Figure 3.6 | The effect on gene expression using RNA extracted in the same experimental conditions, with different commercial RNA extraction kits (Franca, 2011). For a stable gene expression assays, the difference between the cycle threshold of the reference gene and the target gene (Delta CT) should be constant.

A critical view of common RNA quality indicators

The results described on the previous section were somewhat strange, in the sense that we could not predict the outcome of gene expression studies just by determining the quality and quantity of RNA extracted from the biofilms. These indicators are widely accepted as key to determine the success of a proper RNA extraction.

Therefore, we became interested in determining the limitations of these methods in contributing for accurate RNA expression. We envisioned a simple experiment where high quality RNA would be aliquot and then contaminated with either organic components from the biofilm or from organic or inorganic components presented in the RNA extraction kits. Not surprisingly, contaminated RNA had lower A260/A280 or A260/A230 rations (figure 3.7), despite not showing visible signals of degradation, as observed on an agarose gel (data not shown here).

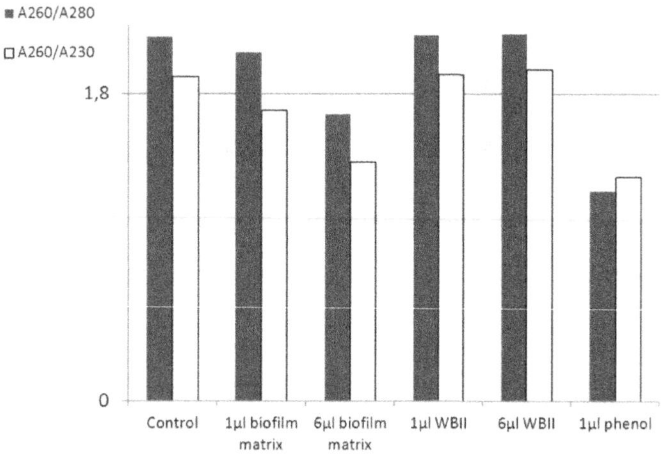

Figure 3.7 | Controlled contamination of high quality RNA and its effect on common quality indicators (Carvalhais, 2013). The same RNA sample was aliquot in six tubes and five were contaminated wither either components of the biofilm matrix or components of the extraction kits used (WBII = Wash Buffer II).

We then determined the gene expression of several independent genes and compared those results with the RNA quality indicators. While low quality RNA clearly influenced the expression of genes, the surprising observation came from RNA contaminated with wash buffer (WBII): despite no significant changes in the RNA quality indicators, gene expression was drastically affected (figure 3.8).

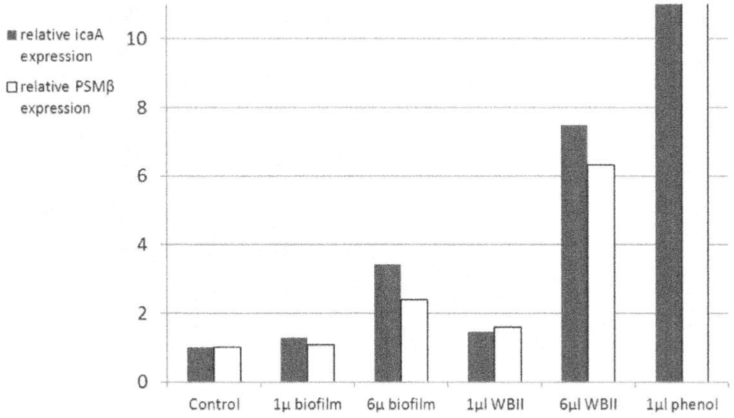

Figure 3.8 | The effect on RNA contamination in gene expression quantification (Carvalhais, 2013). The relative expression of the *icaA* gene was compared between contaminated and clean (control) RNA samples. (WBII = Wash Buffer II).

Furthermore, since RNA contamination did not significantly changed RNA integrity, we performed a similar experiment but, instead of adding contaminants to the RNA aliquots, we performed distinct heat-treatments, in order to induce RNA degradation (Tan, 2009). Interestingly, only drastic heat treatments reduced RNA quality indicators (figure 3.9) but gene expression was still affected in the other situations (figure 3.10). These results clearly confirmed that common RNA quality indicators are not sufficient to guarantee a stable gene expression, further introducing a tricky variable for researchers that are interested in choosing the best set of kits for their experiments. I will get to that point further down in this chapter.

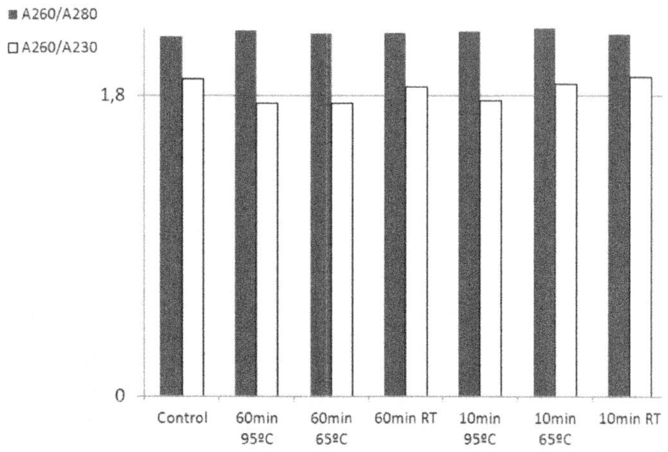

Figure 3.9 | Controlled degradation of high quality RNA and its effect on common quality indicators (Carvalhais, 2013). The same RNA sample was aliquot in seven tubes and six were given distinct heat treatments, in order to induce degradation of the RNA.

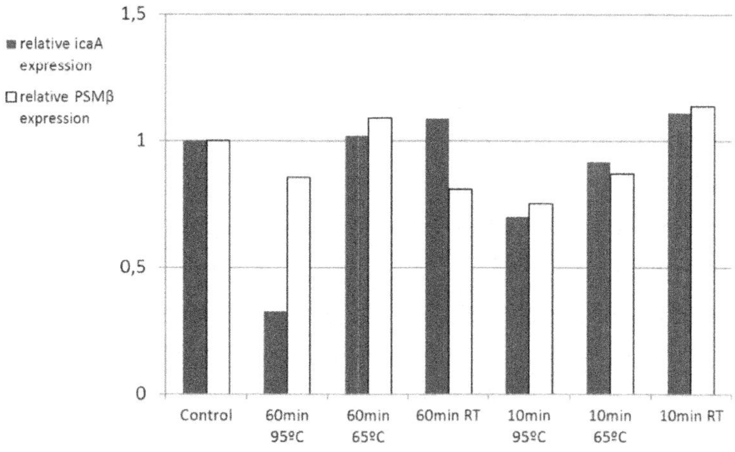

Figure 3.10 | The effect on RNA degradation in gene expression quantification (Carvalhais, 2013). The relative expression of the *icaA* gene was compared between intact (control) and somewhat degraded RNA samples.

The importance of reverse transcriptase and quantitative PCR kits

At this point, it should be already clear that RNA extraction processes have huge impact in the outcome of gene expression studies. This is true not only for biofilm samples, but for many other biological samples, such as blood (Nour, 2010), cancer cells (Tavares, 2011) or even plants (Junttila, 2009), among others. However, gene expression studies involve further experimental steps, besides de RNA extraction. Therefore, it is reasonable to assume that the reverse transcriptase reaction or even the quantitative PCR reaction can also influence the outcome of gene expression studies. A very interesting paper described this effect in eukaryote cells (Sieber, 2010) and we based our next experiments on those findings.

Taking in consideration all our previous work, we extracted RNA of the highest possible quality and then aliquot the RNA prior to perform the subsequent steps. Although we could use a lot more variables, due to financial limitations we selected four distinct reverse transcriptase commercial kits: iScript cDNA synthesis (Bio-Rad), RevertAid First Strand cDNA Synthesis (Fermentas), qScript cDNA Synthesis (Quanta BioSciences) and Super ScriptH VILO (Invitrogen). After RNA was converted to cDNA, by each of the four kits, quantitative PCR was performed also with four different commercial kits: iQ SYBR Green (BioRad), Maxima SYBR Green (Fermentas), PerfeCTA SYBR green (Quanta BioSciences) and mi-real-time EvaGreen (Metabion). Simmilar to the Sieber results, the different combinations of kits tested yielded important differences in gene expression quantification. Interestingly, iScript cDNA synthesis kit was very stable, with any of the tested qPCR kits; conversely, RevertAid demonstrated the highest disparities (figure 3.11).

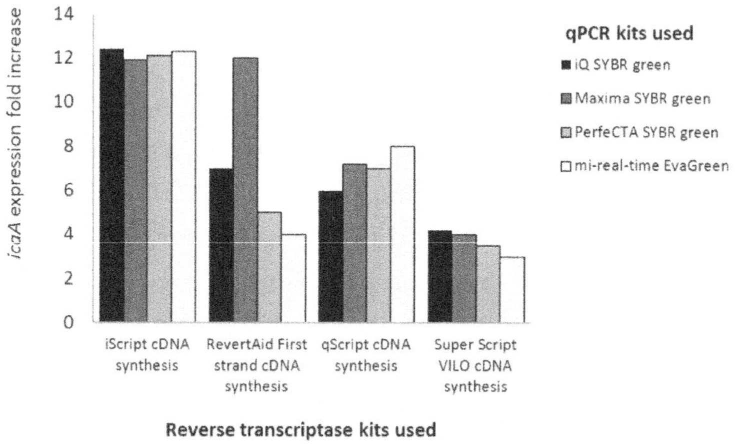

Figure 3.11 | Variability of *icaA* expression due to the usage of different reverse transcriptase and quantitative PCR kits (Franca, 2012b). RNA extracted under the same conditions yielded different quantification of *icaA* expression.

Still, a difficult question remained: should we select the iScript cDNA synthesis kit, due to its lower variability or should we try to understand why Super ScriptH VILO (the second kit with lower variability) yielded the lowest gene expression quantification?

What is the origin of the variability in gene expression studies in biofilms?

As described in the previous sections, the three technical steps involved in common gene expression quantification assays can negatively influence the outcome of our experiments. From this observation, a simple question arose: which of these three was the critical step impairing gene expression reliability?

To try to answer that question, we devised a step-wise approach, in order to quantify the increased variability introduced from each experimental step. We've included not only the three technical steps but also the biofilm formation process, as depicted in figure 3.12.

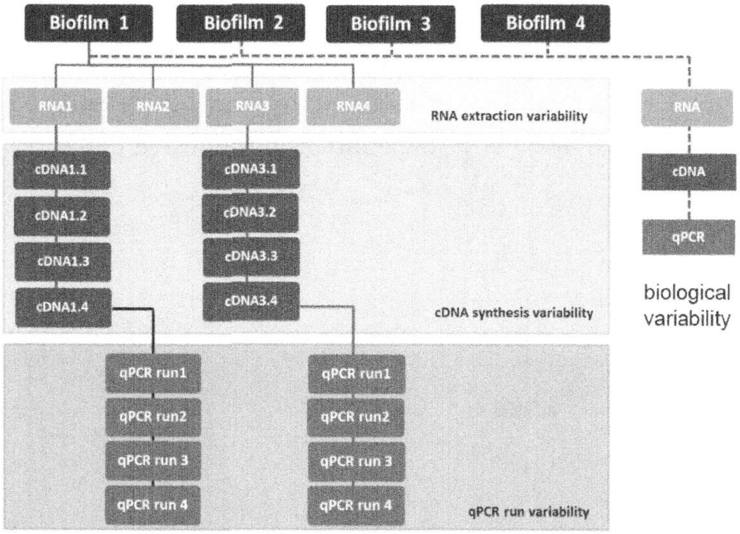

Figure 3.12 | Experimental design to assess the source of gene expression variability in biofilms (Sousa, 2014). One biofilm was divided in four equal shares and four independent RNA extractions were performed. A similar approach was performed from one RNA sample that was used in four reverse transcriptase reactions and latter one cDNA was used in four independent qPCR runs. This process was compared with the traditional workflow: one qPCR run from one cDNA from one RNA from one biofilm.

The expression of five distinct genes was determined: *aap*, which is involved in protein-associated biofilm formation and accumulation (Hussain, 1997); *pgi*, which plays a role in glucose metabolism (Sadykov, 2008); *fmtC*, which is involved in L-lysine modification of phosphatidylglycerol and immune evasion (Staubitz, 2004); *lrgB*, which has been linked with programmed cell death (Sadykov, 2012) and, finally, the biofilm intercellular adhesin *icaA*. Interestingly, biofilm growth was the main variable introducing a higher dispersion of gene quantification (figure 3.13), resulting in lower experimental accuracy. No significant differences were found between RNA extraction and cDNA synthesis. Of note, the variability introduced by the qPCR run alone could reach up to 20%, depending on the tested gene.

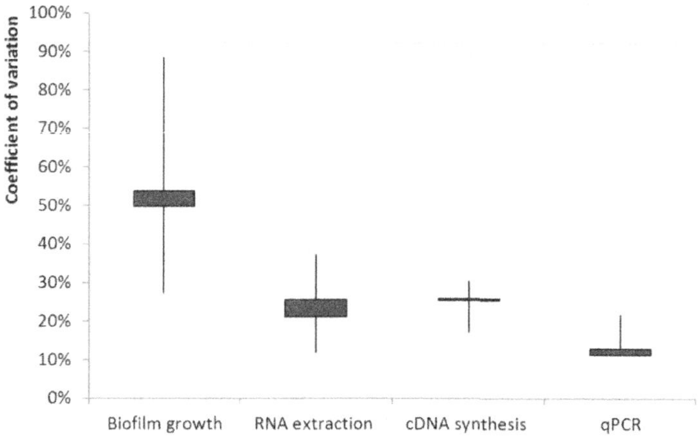

Figure 3.13 | Coefficient of variation of the normalized expression of 5 distinct *S. epidermidis* genes, using the experimental design described in figure 3.12. (Sousa, 2014). For all tested genes, biofilm growth was, by far, the process that introduced higher variability.

Reducing experimental variability

I have to say that the results obtained in the experiments described on the previous section were not surprising: having in mind that biofilms are in fact very heterogeneous samples (Stewart, 2008), one could anticipate that, in fact, the biological variability would account for a significant part of the overall coefficient of variation of the experiment. With this in mind, we then devised a simple experiment: by pooling together up to 20 independent biofilms, per RNA extraction, we hoped to significantly reduce the variability previously observed. Amazingly, with this very intuitive and simple experiment, we solved an issue affecting all researchers working in gene expression in biofilms (figure 3.14). Clearly, different biofilms induced different gene expression outcomes, as we had shown early in this chapter, but the pooling strategy was efficient, independently of the strain tested.

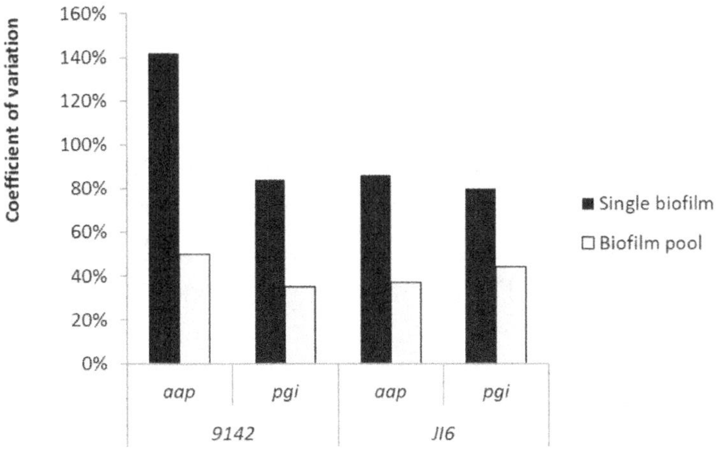

Figure 3.14 | Effect of biofilm pooling in the reduction of the coefficient of variation (Sousa, 2014). This experiment was performed for two independent strains and two distinct genes.

Conclusions

And so, we arrived to the end of this chapter and a 4 year journey into the experimental optimization of gene expression studies in biofilm cultures. For those who were expecting to find the answer to the riddle of choosing proper kits, I'm sorry to disappoint you: there is no such thing (or at least, I do now know what it is). But, the most interesting message that I hope this chapter transmitted to you is that gene expression studies are affected by many intrinsic and extrinsic factors and, ultimately, the detection of one gene up, or down regulated, should not be straightforward associated with an important biological response from the bacteria at study. You need to be absolutely convinced that the phenomenon observed is, in fact, the result of a biological response, and not an artifact of the experimental methodology. And, the only way you should be definitively convinced, is if you excluded all other possibilities... Or at least, try to exclude the ones I've shown you throughout this chapter.

Ackowlegements

The work described in this chapter was only possible due to the research efforts of Ângela França, Virgínia Carvalhais, Cármen Sousa, Luís Melo, Maria Delgado-Rastrollo, Ana Freitas, Ana Henriques and Joana Bento.

References

Carvalhais V, Delgado-Rastrollo M *et al.* (2013) "Controlled RNA contamination and degradation and its impact on qPCR gene expression in S. epidermidis biofilms" in **Journal of Microbiological Methods**, 95:195-200

Cerca N, Brooks JL *et al.* (2008) "Regulation of the intercellular adhesin locus regulator (icaR) by SarA, sigmaB, and IcaR in Staphylococcus aureus" in **Journal of Bacteriology**, 190:6530-6533

Cerca N, Jefferson K *et al.* (2006) "Comparative antibody-mediated phagocytosis of *Staphylococcus epidermidis* cells grown in a biofilm or in the planktonic state" in **Infection and Immunity**, 74:4849-4855

Cerca N and Jefferson KK (2008b) "Effect of growth conditions on poly-N-acetylglucosamine expression and biofilm formation in Escherichia coli" in **FEMS Microbiology Letters**, 283:36-41

Cerca N, Pier GB *et al.* (2005) "Quantitative analysis of adhesion and biofilm formation on hydrophilic and hydrophobic surfaces of clinical isolates of *Staphylococcus epidermidis*" in **Research in Microbiology**, 156:506-514

Franca A, Bento J *et al.* (2012a) "Variability of gene expression in biofilms of foodborne pathogens using RNA extracted with different commercially available kits" in **Current Microbiology**,

Franca A, Freitas AI *et al.* (2012b) "Optimizing a qPCR gene expression quantification assay for S. epidermidis biofilms: a comparison between commercial kits and a customized protocol" in **PLoS One**, 7:e37480-

Franca A, Melo LD *et al.* (2011) "Comparison of RNA extraction methods from biofilm samples of Staphylococcus epidermidis" in **BMC Research Notes**, 4:572-

Gelfand DH and Rudo N (1977) "Mapping of the aspartate and aromatic amino acid aminotransferase genes tyrB and aspC" in **Journal of Bacteriology**, 130:441-444

Heilmann C, Schweitzer O *et al.* (1996) "Molecular basis of intercellular adhesion in the biofilm-forming *Staphylococcus epidermidis*" in **Molecular Microbiology**, 20:1083-1091

Hussain M, Herrmann M *et al.* (1997) "A 140-kilodalton extracellular protein is essential for the accumulation of *Staphylococcus epidermidis* strains on surfaces" in **Infection and Immunity**, 65:519-524

Junttila S, Lim KJ *et al.* (2009) "Optimization and comparison of different methods for RNA isolation for cDNA library construction from the reindeer lichen Cladonia rangiferina" in **BMC Research Notes**, 2:204-

Nour AM, Barbour EK *et al.* (2010) "Comparison of five RNA extraction methods from rabbit's blood" in **Agriculture and Biology Hournal of North America**, 1:448-450

Rodrigues D, Cerca N *et al.* (2011) "*Listeria monocytogenes* and *Salmonella enterica* Enteritidis biofilms susceptibility to different disinfectants and stress response and virulence gene expression of surviving cells" in **Microbial Drug Resistance**,

Sadykov MR and Bayles KW (2012) "The control of death and lysis in staphylococcal biofilms: a coordination of physiological signals" in **Current Opinion in Microbiology**, 15:211-215

Sadykov MR, Olson ME *et al.* (2008) "Tricarboxylic acid cycle-dependent regulation of *Staphylococcus epidermidis* polysaccharide intercellular adhesin synthesis" in **Journal of Bacteriology**, 190:7621-7632

Sieber MW, Recknagel P *et al.* (2010) "Substantial performance discrepancies among commercially available kits for reverse transcription quantitative polymerase chain reaction: a systematic comparative investigator-driven approach" in **Analytical Biochemistry**, 401:303-311

Sousa C, Franca A *et al.* (2014) "Assessing and reducing sources of gene expression variability in *S. epidermidis* biofilms" in **Biotechniques**, 57:1-5

Staubitz P, Neumann H *et al.* (2004) "MprF-mediated biosynthesis of lysylphosphatidylglycerol, an important determinant in staphylococcal defensin resistance" in **FEMS Microbiology Letters**, 231:67-71

Stewart PS and Franklin MJ (2008) "Physiological heterogeneity in biofilms" in **Nature Reviews in Microbiology**, 6:199-210

Tan SC and Yiap BC (2009) "DNA, RNA, and protein extraction: the past and the present" in **Journal of Biomedicine and Biotechnology**, 2009:574398-

Tavares L, Alves PM *et al.* (2011) "Comparison of different methods for DNA-free RNA isolation from SK-N-MC neuroblastoma" in **BMC Research Notes**, 1:140-

Chapter 4

Methods for gene expression studies in biofilms

Ângela França* and Nuno Cerca
CEB - Centre of Biological Engineering, LIBRO - Laboratory of Research in Biofilms Rosário Oliveira, University of Minho, Campus de Gualtar, 4710-057 Braga, Portugal
*** afranca@deb.uminho.pt**

Chapter 4. Methods for gene expression studies in biofilms

Contents of chapter

Introduction to the chapter

As you will learn in the second section of this book, the quantification of mRNA transcripts can be very useful in studying biofilm samples. However, these measurements can be fraught with bias, and high variability can be frequently observed between experiments, resulting in inaccurate quantification and misrepresentative results, as discussed in the previous chapter.

The workflow of gene expression quantification assays by qPCR includes: 1) collection and immediate stabilization of biofilm samples, 2) RNA extraction, 3) complementary (c) DNA synthesis and finally, 4) amplification and quantification of cDNA synthesized molecules. Each of these experimental steps will introduce variability in gene expression quantification. In addition to RNA extraction methods (Franca, 2011), cDNA synthesis and qPCR kits (Franca, 2012b; Sieber, 2010), biological factors such as RNA degradation (reviewed in Fleige and Pfaffl, 2006) and presence of inhibitors (Carvalhais, 2013), are a common cause of inter-experiments variability. It is, therefore, important to be aware of the good practices in order to obtain pure and intact RNA for downstream applications.

Hence, in this chapter, an optimized protocol for gene expression quantification of biofilm samples, by qPCR, will be described. We will go through each step involved in gene expression quantification assays, scrutinizing the several options you have for each step and giving you some important hints to improve RNA quality and thus, the accuracy and reliability of your gene expression quantification assays. As described in the previous chapter, biofilm growth will introduce high variability and we suggest you to pool several independent biofilms for each RNA extraction.

RNA extraction

Nowadays, there are several commercially available kits to isolate total bacterial RNA. However, different kits can give you varying levels of RNA quality (Franca, 2011; Franca, 2012a). This will have direct impact on gene expression quantification accuracy as any RNA-based analysis depends on the quantity, purity and integrity of the RNA extracted (reviewed in Fleige and Pfaffl, 2006). As shown in figure 4.1, RNA quality can be influenced by several factors such as sample collection (Qiagen,RNeasy mini kit handbook), ribonucleases activity (Life Technologies, Top 10 ways to improve your RNA isolation), RNA extraction methods (Franca, 2011) and storage conditions (Schoor, 2003).

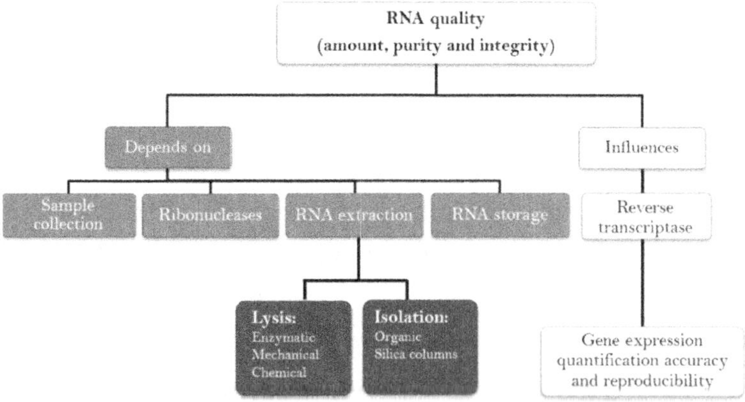

Figure 4.1 | Factors influencing RNA quality. RNA quality can be impaired during extraction or storage.

Sample collection

Although all the above mentioned factors have an important impact in RNA quality, the conditions of the starting material are of upmost importance, in order to have high quality RNA. Right after harvesting your samples, changes in the gene expression profile can occur due to specific (enzymatic) and non-specific (mRNA lifespan) RNA degradation or due to transcriptional induction (Qiagen, RNeasy mini kit handbook). Therefore, to avoid unspecific alterations in gene expression, an immediate stabilization of the biological sample is essential for reliable gene expression quantification. There are several commercially available solutions, such as RNAlater® (for tissues, Ambion) and RNAprotect™ bacteria reagent (for bacteria, Qiagen), which are able to protect RNA from degradation by specific or non-specific processes. These reagents should be used whenever conceivable, and need to be added as soon as possible during sample collection procedure.

Ribonucleases

After sample collection, the introduction of external ribonucleases, vulgarly known as RNases, is an important cause of RNA degradation. RNases are ubiquitous enzymes that catalyze the degradation of RNA molecules and are found in almost every cell types and organisms, playing essential roles in nucleic acid metabolism, as well as in protection (Cheng and Deutscher, 2005; reviewed in Matos, 2014). To increase the challenging associated with their ubiquity, RNases are highly specific. Even small quantities of these enzymes are able to degrade high amounts of RNA (Bioline, Working with RNA: hints and tips), especially if you are working with low concentrated samples. Furthermore, RNases are highly stable enzymes that can regain part of their activity after long periods of boiling or autoclaving (Life Technologies, The Basics: RNase control). It is, thus, important to follow some basic precautions to decrease contamination by exogenous RNases. The most important conducts are listed below:

- Avoid extracting RNA on places with open windows or next to the door or air flows;
- Clean your bench with water (to remove durst), alcohol (to remove microorganisms) and a solution capable of denaturing RNases such RNase AWAY™ or a solution containing 0.1 M NaOH and 1 mM EDTA;
- Clean your set of micropipettes with RNase denaturing solution. If possible spare one set of micropipettes only for RNA handling;
- Use tubes, tips, water, buffers and solutions RNase-free certified or, when possible, perform diethylpyrocarbonate (DEPC) treatment;
- Wear gloves and change it frequently throughout the experiment. Do not forget to change gloves every time you touch skin, hair, doorknobs, keyboards, etc;
- Immediately after use, put the lid back in all reagents to avoid contamination with RNases. Alternatively, you can perform your experiments under a protected place or aliquot reagents for each extraction handling stock reagents only inside a clean laminar flow cabinet.

RNA extraction kits

As described on the previous chapter, the principle used by the RNA extraction kit selected for your experiments is another aspect influencing RNA quality (Franca, 2012a; Rump, 2010). Currently, there are several commercially available kits for RNA isolation and, depending on the lyses method (chemical, enzymatic or mechanical), as well as on the isolation principle applied (organic or columns), mRNA quality can be impaired (Franca, 2011). The nature of your samples such as Gram-positive or Gram-negative structures, as well as the phenotype, planktonic or biofilm, will also impact the efficacy of the kit used. It would be, therefore, advisable to test different RNA extraction kits before starting your experiments, in order to select the most appropriated kit: the kit that yields the highest concentration of the purest and intact RNA.

RNA storage

After RNA isolation, another important aspect that will affect RNA integrity is its storage. Samples that are stored under sub-optimal conditions are more susceptible to degradation (Schoor, 2003). Ideally, RNA should be eluted, or suspended, in nuclease-free water supplemented with 0.1 mM EDTA or in Tris-EDTA buffer, due to their chelating functions, and stored at -20 ºC for short term storage (1 month) or, for long term storage, you can store your RNA samples at -80 ºC (1 or more years) or as ethanol precipitates at -20 ºC (Life Technologies, Top10 ways to improve your RNA isolation). Nevertheless, if you have handled your RNA with all precautions, RNA can be stored in nuclease-free water only, maintaining its integrity for over a year if stored at -80 ºC. If possible, making aliquots will avoid freeze-thaw cycles that may result in RNA unspecific breakage and will also reduce the probability of introducing RNases (Life Technologies, Top10 ways to improve your RNA isolation).

Experimental protocol

This RNA extraction protocol was optimized for biofilm samples analysis. It combines the most efficient lysis method for Gram-positive bacteria, mechanical lysis, with the faster isolation procedure of the market, silica membranes. As you can see in chapter 3, several silica membrane kits may be used in this combined protocol. Herein, we have used the kit RNeasy mini (Qiagen). Keep in mind that the name of the buffers, or volumes, presented here may be different from the kit you will select for your assays. To start mRNA transcripts isolation, please follow the indications below.

- Remove spent medium from established biofilms;
- Carefully wash biofilms once with phosphate buffered saline;
- Immediately add 1 mL of RNAprotect™ bacteria reagent (or similar solution) diluted 2:1 in nuclease-free water or phosphate buffered saline;

! be aware ! If you will used RNA protect solutions, you can safely stop the procedure right here. Depending on the brand and particularities of the solution used, you may leave the samples for 1 day at 37 °C, 1 week at 4 °C and several months at -80 °C, without deteriorating RNA quality.

- Collect biofilm cells by scraping then from the plate with a tip or 1mL insulin syringe embolus;

- Pool ≥ 10 biofilm together (as described in the previous chapter);
- Collect 1-2 mL of the pooled biofilm suspension and centrifuge at 5000 *g* for 10 minutes at 4 °C;

! be aware ! If using other stabilizing solution than RNAprotect™ bacterial reagent (Qiagen), please see manufacturer's recommendations for centrifugation.

- Discard the supernatant and suspend the pellet in 500 µL of lysis buffer (RLT), supplemented with β-mercaptoethanol, and 500 µL of molecular grade phenol;

! be aware ! Although the supplementation with β-mercaptoethanol is optional, due to its RNase inhibitor properties, we do advisable you to use it.

! be aware ! Do not forget that **β-mercaptoethanol** is considered toxic, causing irritation to the nasal passageways and respiratory tract upon inhalation and also irritation to the skin. It can cause vomiting and stomach pain through ingestion. **Phenol** is irritating and corrosive to the skin being rapidly absorbed, whereupon it can affect the central nervous system and cause damage to the liver and kidneys. It can also cause severe damage to eyes, including blindness. Phenol is also a mutagen. **Please use adequate safety equipment when handling these samples, and always use a fume hood.**

- Transfer the bacterial suspension into tubes with glass beads;

! be aware ! The size of the beads will be different depending on the microorganisms the will be used. For bacteria, beads should have size between 150 to 210 µm. For yeasts, beads should have size between 400 to 500 µm.

- Using a cell disruptor, lyse biofilm cells at 6.5m/s during 35 seconds;
- Repeat this step three times with 5 minutes intervals on ice;

! be aware ! This protocol was optimized for gram-positive bacterial biofilms. If using other biofilms, the number of cycles needs to be optimized.

- Centrifuge the lysate at 16000 *g* for 2 minutes;
- Transfer the supernatant into a new RNase-free tube. It is better to leave behind a small amount of suspension to avoid bringing beads, lowering the quality of the RNA;
- Add 500 µL of 70% ethanol and mix by inverting the tubes;

! be aware ! The addition of ethanol is very important to allow the selective binding of RNA molecules to the silica membrane of the purification columns. Moreover, ethanol should be freshly prepared. 70% ethanol tends to absorb water from the air and hence, should be prepared at the moment for optimal results. Do not forget to use RNase-free water for ethanol dilution.

- Transfer 700 µL of the lysate into the RNA isolation column;

! be aware ! The maximum capacity of these columns is 700 µL. Hence, if you have more volume you will have to transfer 700 µL at each time, until all your lysate have passed through the column. Nevertheless, depending on the kit used this volume can be different.

- Centrifuge the column at ≥ 10000 *g* for 30 seconds;

- Discard the flow-through and transfer the remaining lysate into the spin column. Reuse the collection tube;

! be aware ! All RNA molecules will be bound to the silica membrane in this step. Undesirable molecules and potentially inhibitors will be washed away.

- Add 700 µL of wash buffer 1 – RW1;
- Centrifuge at ≥ 10000 *g* during 30 seconds;
- Discard the flow-through and place the spin column back into the collection tube;
- Add 500 µL of wash buffer 2 – RPE;
- Centrifuge at ≥ 10000 *g* during 30 seconds;
- Discard the flow-through and place the spin column in the same collection tube;
- Add 500 µL of wash buffer 2 – RPE for the second time;
- Centrifuge at ≥ 10000 *g* during 2 minutes;
- Discard the flow-through and place the spin column into a new collection tube;
- Centrifuge at ≥ 10000 *g* during 3 minutes to remove any traces of ethanol;

! be aware ! Wash buffer 2, in this case RPE, is manly composed by ethanol. Ethanol carry-over will impact downstream applications and is why this additional 3 minutes centrifugation is needed.

- Transfer the spin column into a 1.5 mL RNase-free tube;
- Add 50 µL of nuclease-free water into the center of the silica membrane;

! be aware ! Ensure that the elution buffer/water is dispensed directly into the center of the silica membrane. If the elution buffer/water is lost within the walls of the column, RNA yield will be lower.

- Incubate for 1 minute at room temperature;
- Centrifuge the column at full speed for 1 minute;
- Immediately transfer RNA to ice.

Genomic DNA removal

Despite the majority of the kits used for RNA extraction are very specific, it is virtually impossible to have a sample completely free of genomic (g) DNA (Life Technologies, Now it's easy to make your RNA free of genomic DNA contamination and ready for RT-PCR). However, as you can imagine, the presence of gDNA has important repercussions in gene expression quantification, as it will result in the detection of false positives and considerably influence the analysis of low expressed genes (Life Technologies, Now it's easy to make your RNA free of genomic DNA contamination and ready for RT-PCR). Furthermore, the presence of gDNA, depending on the method used for RNA concentration determination, can overestimate the amount of total RNA, having significant effects in downstream applications. Hence, the decreased of genomic DNA to insignificant levels, is required for sensitive downstream applications such as qPCR, RNA-sequencing or microarrays. Nevertheless, the level of gDNA contamination can be determined during reverse transcription reaction (for more information see the section "controls").

There are 2 major methods to decrease the amount of gDNA. You can either use organic methods, such as phenol-chloroform procedure and lithium chloride precipitation, or degrade it using specific enzymes such as DNase. Most commonly, DNase will be used. Importantly, when using DNase, you always need to efficiently inactivate the enzyme at the end of the protocol because, although with fewer efficacies, DNase can also degrade mRNA:DNA hybrid molecules, such as the ones that will be synthesized in the first strand cDNA synthesis, or even degrade double stranded DNA products of the qPCR run (Life Technologies, DNase I demystified).

DNase inactivation can be performed by 4 different ways: (i) heating to 65 ºC, (ii) phenol extraction, (iii) using a highly dense reagent that binds to DNase, as well as divalent cations, and precipitate them at the bottom of the tube, and (iv) by subsequent washing steps in DNase in-column procedure. Differently from the inactivation by heat, the last three methodologies do not promote RNA degradation. Yet, the most used so far is the inactivation by heat. Since RNA is very unstable when in the presence of divalent cations such as Ca^{2+}, and Mg^{2+}, at temperatures higher than 65 ºC, DNase inactivation is usually performed in the presence of an effective chelating agent such EDTA. EDTA binds these cations protecting RNA from heat-induced strand scission.

In this protocol RNA-free DNase I (Thermo Scientific, MA, USA) treatment was used, followed by enzyme inactivation by heat. The procedure is as follows:

- Add 5 uL of DNase buffer and 2.5 uL of DNase I;
- Mix thoroughly performing up-and-down;
- Incubate the tubes for 30 minutes at 37 °C;
- Add 5 uL of EDTA and mix thoroughly performing up-and-down;
- Incubated for 10 minutes at 65 °C.

! be aware ! RNA may be directly used for downstream applications or safely stored until further use.

RNA concentration and purity

Although the recovery of low amounts of RNA may not seem an important issue, the use of low quantity of RNA for downstream applications may place low expressed genes below the limit of detection. Thus, the use of low concentrated samples can result in either non-reproducible or non-accurate results (Sieber, 2010).

RNA concentration is traditionally measured using UV spectroscopy, typically by Nanodrop™, due to the low sample consumption (as low as 1.0 µL) and easiness of aplication. The concentration is then determined using Beer-Lambert law, knowing that an absorbance reading, at 260 nm wavelength, of 1.0, is approximately 40 µg/mL for single stranded RNA and 50 µg/mL for double stranded DNA (Promega, Calculating nucleic acid or protein concentration). However, UV-based techniques cannot distinguish between nucleic acids from free nucleotides, or even DNA from RNA molecules (Jones, 1998). Moreover, some contaminating compounds absorb close to 260 nm, what can overestimate the concentration values determined (Fleige and Pfaffl, 2006). Alternatively, RNA concentration can be accurately determined, among others (Life Technologies, Nucleic acid quantification in solution), using fluorescent dyes such as RiboGreen® (Jones, 1998). This dye is able to distinguish nucleic acids from free nucleotides and DNA from RNA molecules, and is not influenced by the presence of common sample contaminants (Jones, 1998). However, in this case, no information regarding sample purity is given.

Nevertheless, Nanodrop™ is frequently used to assess RNA purity through the measurement of both 280 and 230 nm wavelengths (NanoDrop, 260/280 and 260/230 ratios). The presence of impurities such as phenol and/ or chaotropic salts and biofilm matrix components is indicated by the ratio A_{260}/A_{230}. These compounds are generally introduced by the lysis buffers used in RNA extraction kits in order to protect RNA from endogenous and exogenous RNases activity, or due to the presence of biofilm matrix, respectively The presence of these compounds have inhibitory effects on cDNA synthesis and, consequently, in qPCR run (Carvalhais, 2013). The same occurs for protein contamination, or some phenolic compounds (NanoDrop, 260/280 and 260/230 ratios), which absorb preferentially at 280 nm and is, thus, evaluated by the A_{260}/A_{280} ratio. It was established that pure RNA should present both A_{260}/A_{280} and A_{260}/A_{230} ratios higher than 1.8, but, ideally, these ratios should be above 2.0. Because of the significance of such indicators, it is important to be aware that these purity values are highly dependent on pH and ionic strength of the diluent used (Wilfinger, 1997). As can be seen in table 4.1, it is advisable to use low salt buffers such as Tris or Tris-EDTA to accurately determine these purity indicators (Life Technologies, The basics: RNA isolation).

Table 4.1 | Influence of pH and ionic strength on A_{260}/A_{280} ratio values. Adapted from www.lifetechnologies.com.

Eluent/diluent (pH)	A_{260}/A_{280}
DEPC-treated water (pH 5-6)	1.6
Nuclease-free water (pH 6-7)	1.85
Tris-EDTA buffer (pH 8.0)	2.14

If your samples have low purity, you may use commercially available kits to purify RNA already isolated, or you can also try ethanol, lithium (specific for RNA) or ammonium acetate precipitation protocols. Alternatively, if your target gene is being expressed in a satisfactory amount, the best way to get rid of impurities is to dilute the sample until no influence on gene expression quantification is noticeable (Franca, 2012b).

Experimental protocol

In this particular protocol, RNA concentration and purity will be assessed by Nanodrop™1000. Follow the specifications below to ensure correct RNA concentration determination.

- Carefully clean the NanoDrop™ pedestal with water;
- Add 5 μL of water into the pedestal and initialize the spectrophotometer;
- Let it warm up for 10 minutes;

- Apply between 2 and 3 μL of the elution buffer/ water in the pedestal;
- Define the blank;

! be aware ! Make sure that the buffer used to set the blank is the same used for RNA suspension or elution. Otherwise concentration values will not be accurately calculated.

- Confirm that your blank is really zero;
- In "sample type" option select RNA;

! be aware ! It is very important to select "RNA" in sample type in order to accurately quantify RNA concentration. Remember that A_{260nm}=1 is equivalent to 40 μg/mL of single stranded RNA, while it corresponds to 50 μg/mL of double stranded DNA.

- Apply 2.5 μL of your sample in the pedestal;

! be aware ! The minimum that you can use is around 1.5 μL. However, if you can use more, like 2-3 μL, use it since it may avoid errors associated with low volume.

- Measure both concentration (ng/μL) and purity indicators (A_{260}/A_{280} and A_{260}/A_{230});

! be aware ! Remember that both ratios should be higher than 2, or at least higher than 1.8.

- At the end confirm blank value once more. If you have too many samples to read, please ensure that blank is correct in the middle of the measurements as well;
- Wash the pedestal by placing 5 μL of water in the pedestal and click measure. This way water go through the entire internal system and cleaning it;

- Repeat this step twice;
- Switch the equipment off.

RNA integrity

RNA integrity is another important quality indicator, if not the most important of all. Several studies have shown that RNA integrity impair quantification of mRNA molecules (reviewed in Fleige and Pfaffl, 2006) being, therefore, essential to evaluate its integrity before proceed to any downstream application.

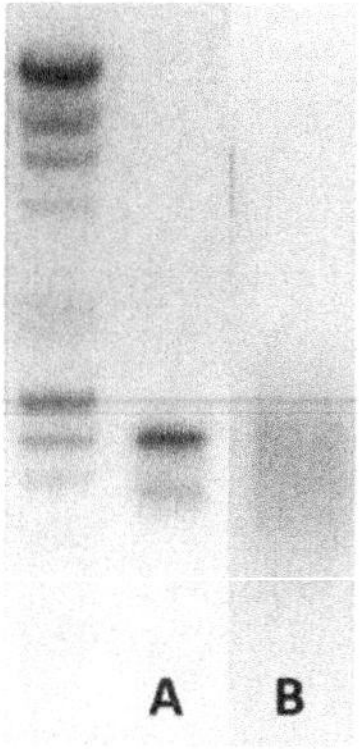

Figure 4.2 | Example of a non-denaturant agarose gel (1.5%) showing intact (A) or degraded (B) total RNA.

The simplest method to assess RNA integrity is by loading the sample in an agarose gel, either denaturant or non-denaturant. It is important to note that only rRNA will be observed. Intact rRNA will show 2 sharp bands, without smear, corresponding to 23S and 16S rRNA subunits (in the case of bacteria). Furthermore, another indicator of integrity is the intensity of each band. For samples with intact RNA, 23S rRNA band has to be as twice as much intense than 16S rRNA (Pinto, 2009). An example of intact or degraded RNA can be seen in the figure 4.2, lane A and B, respectively.

Despite its easiness, this method is becoming obsolete due to its subjectivity of analysis and the need of high amounts of RNA for visualization. In its place, lab-on-a-chip approaches, a combination of microfluidics, capillary electrophoresis and fluorescent dyes, have been widely used to evaluate RNA integrity but also, RNA concentration (Schroeder, 2006). The most used instruments are the Bioanalyzer 2100, from Agilent Technologies (CA, USA) and Experion™, from Bio-Rad (CA, USA). Lab-on-a-chip analyses calculate a quality value, known as RNA integrity number (RIN) (Bioanalyzer) or RNA quality indicator (RQI) (Bio-Rad), that is based on the entire RNA electrophoretic trace and not only on the peaks of the two major ribosomal subunits (Schroeder, 2006). The final result is given by electropherograms along with a gel-like image (figure 4.3).

These integrity indicators, which may range from 1 (completely degraded samples) to 10 (intact RNA), allows a more objective interpretation of RNA integrity. Although for prokaryotes it was not yet defined an ideal RIN or RQI values (Bhagwat, 2013), it is accepted, for eukaryotes, that integrity values above 7 are suitable for gene expression analysis (Jahn, 2008). Hence, so far, the same threshold has been used for prokaryotes. RIN or RQI values below 7 may still be used but an extensive validation has to be performed before in order to assess if is still good enough for accurate and reproducible gene expression quantification.

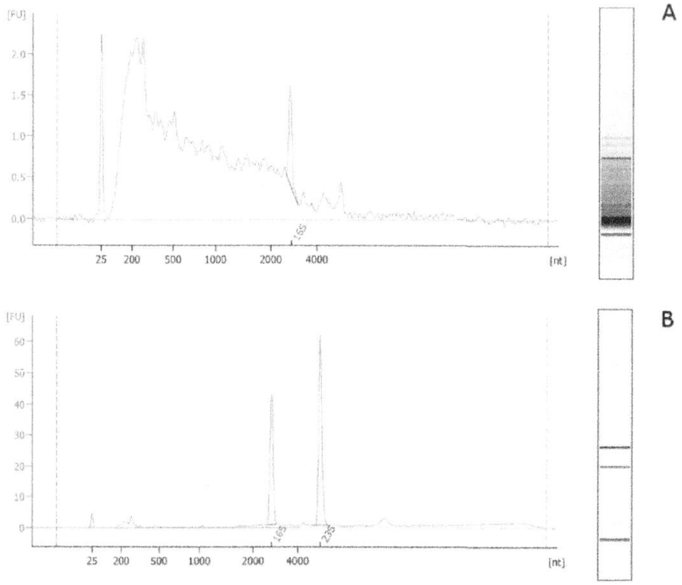

Figure 4.3 | Electropherograms of bacterial RNA samples showing different degrees of degradation as determined by Bioanalyzer 2100. A) Total RNA highly degraded with a RIN=3.0, and B) Total RNA with high quality with a RIN=9.1. FU-fluorescence units; nt-nucleotides.

The verification of RNA integrity before gene expression quantification allows the comparison between experiments and a more objective evaluation of the significance of the results obtained. Henceforth, these lab-on-a-chip technologies are becoming the option of choice, particularly for gene expression profiling assays.

Experimental protocol

Although agarose gels present some limitations, as described above, it is still a widely used technique for a quick evaluation of RNA integrity. Hence, before performing a lab-on-a-chip technology, which is expensive, it could be interesting to run an electrophoresis to previously select your samples, or to have an idea of your extraction performance. However, it is important to keep in mind that it is very difficult to have RNase-free conditions when run a gel and thus, some observed degradation may occurred during the run and not a result of your RNA extraction procedure. Before starting, clean all material needed for electrophoresis, including tanks, trays and the comb following the specifications below:

- Wash the electrophoresis material with 0.1% SDS or 0.1 M of NaOH and 1M EDTA solution;
- Rinse with RNase-free water;
- Rinse with absolute ethanol;
- Rinse, thoroughly, with RNase-free water an protect all material from dust;
- Weight 1g of agarose and mix with 100 mL of 1× Tris-Acetate-EDTA (TAE) or Tris-Borate-EDTA (TBE) buffer;
- Let agarose cool down for 10 minutes;
- Add nucleic acid staining solution;
- Pour the gel and let it solidify (protect the gel from dust and light);
- If possible, normalize the concentration of your samples to 1500 ng of total RNA in nuclease free water;
- Heat the sample for 5 minutes at 70 °C;
- Place the samples immediately on ice;

! be aware ! This heat-cool cycle will break RNA secondary structures what will linearize the molecule and will improve its electrophoretic mobility. This step is important if you want to accurately determine the length of the ribosomal subunits. If is only to visualize integrity this step is not essential.

- Add the loading dye to your samples and load them into the gel;
- Run the gel at 80 Volts for 40-60 minutes.

At the end you should obtain an image like figure 4.2A. Otherwise careful should be taken when using that sample.

Complementary DNA synthesis

Complementary DNA (cDNA) is a double stranded hybrid molecule synthesized from mRNA molecules. This reaction is catalyzed by the reverse transcriptase enzyme, which uses the strand of mRNA as a template to construct its complementary DNA strand, based on the pairing of RNA base pairs to their DNA complements.

There are two different ways to synthesize cDNA, which depends in the qPCR approach that will be used. qPCR can be performed directly using RNA, being both reverse transcriptase and cDNA amplification and quantification reactions completed in the same tube. This approach is known as one-step PCR (reviewed in Wong and Medrano, 2005). On the other hand, gene expression quantification can be performed using previously synthesized cDNA, being reverse transcriptase and qPCR completed in two different tubes. This approach is known as two-step qPCR (reviewed in Wong and Medrano, 2005). The best approach to be used is still controversial (Wacker and Godard, 2005) and we advise you to test different kits in order to validate your system.

One-step qPCR is characterized by its convenience: it is less time-consuming and requires less user intervention than two-step qPCR, minimizing the chances of pipetting errors and contamination (reviewed in Wong and Medrano, 2005) Nevertheless, there is some controversy between companies and researchers. Contrary to what is defended by manufacturers, some studies have shown that one-step reaction is intrinsically variable and significantly less sensitive and reproducible than two-step qPCR system (Battaglia, 1998; Vandesompele, 2002a). Furthermore, it is not possible to store cDNA for any further analyses when performing one-step qPCR. Two-step qPCR, on the other hand, is characterized by its flexibility, possibility of optimization, as well as the possibility to store cDNA for any subsequent analysis, being ideal for difficult to obtain samples as it allows quantifying several genes using the sample (Wacker and Godard, 2005).

Priming strategy

Another issue that needs to be addressed, before starting cDNA synthesis, is the priming strategy that should be used. It is important to remember that RNA degradation, as well as the presence of impurities, has a direct impact on the percentage of target mRNA molecules that are converted into cDNA (Zhang and Byrne, 1999). However, depending on the first-strand priming strategy used, this issue can be minimized.

There are 3 different types priming: (i) oligo (dT), which are only used for polyadenylated RNA (eukaryotes organisms), (ii) random sequence oligonucleotides and (iii) sequence specific primers, which are particularly interesting in the case of non-polyadenylated RNA, such as bacterial RNA (Life Technologies, RNA priming strategies).

Random sequence oligonucleotides, usually short segments ranging from 6 to 9 nucleotides long, consisting of every possible combination of bases. This is the most flexible priming strategy as it creates a large pool of cDNA and enables you to use the same sample to quantify several different genes. Furthermore, as these primers anneal throughout the target molecule, the concern of primer binding efficiency, due to RNA degradation or presence of secondary structures, is partially resolved (Life Technologies, RNA priming strategies). When template is limiting, cDNA synthesis with random sequence oligonucleotides may yield slightly higher number molecules than sequence-specific priming (Thermo Scientific, Basic principles of RT-qPCR). It is important to be aware that this priming strategy can however overestimate the number of cDNA molecules copies (reviewed in Bustin, 2005) and, since it binds to all RNA (tRNA; rRNA and mRNA), it can dilute mRNA signal (Thermo Scientific, Basic principles of RT-qPCR).

Oppositely to random sequence oligonucleotides, in sequence-specific priming strategy, only target genes are retro-transcribed. This priming strategy offers the highest specificity and consistency of all options. Nonetheless, secondary structures along with RNA degradation, can highly impact primer binding ultimately impairing cDNA synthesis efficiency and qPCR run results (Life Technologies, RNA priming strategies). Furthermore, cDNA synthesized using sequence-specific priming is less flexible than cDNA synthesized using random sequence oligonucleotides, as it can only be used for the quantification of the genes which primers you used, and no other genes that you may be interested to study later on (Eurogentec, qPCR guide).

Experimental protocol

In this particular experiment, two-step qPCR approach will be used. Hence, cDNA synthesis (Thermo Scientific, CA, USA) will be performed previously to qPCR run and it will be performed using the flexible random sequence oligonucleotides (NZYTech, Lisbon, Portugal) as priming strategy. Please prepare the following:

- In one 0.2 mL RNase-free tube, add 40 to 5000 ng of total RNA;

! be aware ! The minimum and the maximum optimal amount of total RNA that should be used will depend on the reverse transcriptase enzyme selected. Hence, if you are using a different kit, please read the manufacturer's recommendations before. Nevertheless, avoid using less than 40 ng of total RNA as it may result in inconsistent and inaccurate results (Sieber, 2010).

- Add 1 μL of random sequence oligonucleotides or alternatively use 1 μL of 10 mM of reverse sequence-specific primers;

- Add RNase-free water up to 10 µL;
- Mix gently by pipetting up-and-down and spin down the samples;
- Incubate the mixture for 5 minutes at 65 °C
- Chill for 5 minute on ice, spin down the samples and place the tube back on ice;

! be aware ! This heat-cool process is advisable for RNA templates GC-rich sequences or sequences that contain secondary structures.

- Add the following components into a new tube by the order indicated to make the cDNA synthesis master mix:

2 µL of RT buffer (5×) × *n*
1 µL of dNTPs (10 mM each) × *n*
0.75 µL of RevertAid™ H minus RT× *n*
0.25 µL of RiboLock RNase inhibitor × *n*
1 µL of RNase-free water × *n*

! be aware ! *n* is the number of test tubes that you need to use. To avoid sample-to-sample variation you should perform a master mix for all your test samples and controls. Note that in the no reverse transcriptase control (NRT) the reverse transcriptase enzyme is not added and no template is added on the negative control. Due to pipetting errors, we advise you to calculate the volume needed for n+ 1 reactions.

- Transfer 5 µL of previously prepared RNA into a new 0.2 mL RNase-free tube and add 5 µL of cDNA synthesis master mix;
- Using the 5 µL of the RNA that remained for all samples, please add 5uL of a master mix prepared with all

reagents, with exception of the reverse transcriptase enzyme (no reverse transcriptase control). This control will be used for gDNA contamination determination;

- Mix all the contents;
- Spin the samples briefly to avoid bubbles;
- Prepare the following protocol in a thermal cycler:

25 ºC for 10 minutes
42 ºC for 60 minutes
70 ºC for 10 minutes

! be aware ! The 10 minutes incubation at 25 ºC is specific for random sequence oligonucleotides. If you are using other priming strategy please read the manufacturer's instruction first.

! be aware ! cDNA may be immediately used or stored until further use. For short term storage (2 days) cDNA can be placed at 4 ºC and for long term storage it is advisable to store cDNA at -20 or -80 ºC. In these conditions, cDNA integrity will be intact for a long period of time.

Quantitative PCR run

As referred in the chapter 2, qPCR is one of the most sensitive, efficient, fast, accurate and reliable techniques to measure gene expression. However, there are several important considerations, and concepts, that need to be addressed before its use, to ensure reliable analysis and conclusions.

Absolute versus relative quantification

The first question that needs to be asked is if you are interested in determining the exact number of target transcripts, or if you are interested in compare the relative quantity of a transcript between samples. Depending on the answer to these questions, qPCR quantification approach used will be fairly different. There are two principal approaches of quantification: absolute or relative (Pfaffl, 2004).

Absolute quantification is used to determine the exact number of copies of a particular transcript or transcripts. For that, a standard curve, with cDNA or RNA of known concentration, has to be performed by diluting the sample several orders of magnitude, and amplifying it by qPCR (reviewed in Bustin, 2000). The standard curve will produce a linear relationship between the cycle threshold (Ct) obtained and the known concentration (Heid, 1996). Then, your can extrapolate the concentration of other transcripts in your sample based on the Ct obtained (Heid, 1996).

On the other hand, relative quantification gives you the fold change increase/decrease compared with a control sample rather than an exact number of transcripts (Pfaffl, 2004; Pfaffl, 2014). Due to its easiness and accuracy, relative quantification is the most used quantification method. Although this quantification method can also be performed using a standard curve, a normalization approach is generally used (Pfaffl, 2004; Pfaffl, 2014).

Normalization

Normalization is the method whereby the variability resultant from either biological biases or technical variation, introduced throughout the qPCR workflow, can be reduced (reviewed in Bustin, 2005). In order to obtain biologically meaningful results is important to reduce such variability. This reduction can be achieved by standardizing all the steps involved in gene expression workflow.

There are several normalization strategies available; however, the majority of those strategies are not able to completely address the variation associated with each step involved in gene expression quantification workflow (reviewed in Huggett, 2005). These strategies include normalization to the (i) initial sample quantity or tissue volume, (ii) total RNA concentration or to a (iii), reference gene (ribosomal or messenger RNAs) or (iv) alien molecules, or group, of reference genes (reviewed in Fleige and Pfaffl, 2006 and Huggett, 2005). The normalization to the initial quantity consists in the use of the same amount of starting sample for all experiments. Although this can minimize technical variability it does not address the variability associated with RNA isolation and following procedures (reviewed in Huggett, 2005). Furthermore, is not always easy to determine, and normalize, the number of cells between samples. RNA quantity normalization, on the other hand, accounts with the variability associated with RNA isolation. However, it does not include the differences in the efficiencies of following reactions (reviewed in Huggett, 2005).

Finally, the normalization to a reference gene, formerly named as housekeeping gene, is the most complete and the most used normalization method (reviewed in Bustin, 2000). Reference genes are expressed in the same cells of the genes of interest, they go through the same conditions of collection, same isolation procedure, storage and enzymatic treatments, addressing, therefore, almost every source of variability associated with gene expression assays (Bustin, 2009). Nevertheless, care should be taken regarding the level of expression of references genes (reviewed in Thellin, 1999). Reference genes expression has to be constitutive, i.e., unaltered independently of the involving environmental conditions. Small variations may occur under particular conditions and will blur real changes, creating erroneous results and misleading interpretations (reviewed in Bustin, 2000 and Thellin, 1999). Therefore, their expression should be carefully determined under the test conditions before its use.

Due to the difficulties associated with finding reference genes that do not change its expression under test conditions, a more accurate reference gene normalization-based method of normalization was suggested. Vandesompele and his colleagues (Vandesompele, 2002b) proposed that instead of using only one reference gene, at least 3 references genes, with minimal variation across the treatment, should be selected. This method is considered, nowadays, to be the most accurate of all normalization methods. Software such as geNorm, REST, Bestkeeper and NormFinder may be helpful in selecting the most appropriated reference genes for your particular microorganism and experiments.

Controls

The results obtained by qPCR will only be valid if appropriate controls have been included. qPCR controls are divided into 2 main groups: negative and positive controls. Negative controls are performed with the primary objective of detecting contaminations and positive controls with the main goal of determining the quality of the reagents used, the presence of inhibitors and sample quality.

Among the negative controls, no template (NTC) and no reverse transcriptase (NRT) controls are the most important (Eurogentc, qPCR guide). NTC is the control that contains all the reagents of the qPCR run with the exception of the template. This control will indicate the presence of contaminants in your reagents (Qiagen, PCR protocols & applications). In this situation, it is not expected to observe any amplification curve although, sometimes, primer dimer formation in SYBR Green-based chemistry can result in late amplification curves (for more details see "Detection chemistries"). However, as long as the cycle threshold of this curve is 8 cycles a part from the specific signal, no problem with false positives is expected (Eurogentec, qPCR guide).

NRT is prepared during cDNA synthesis. In the NRT tube, every single reagent used for reverse transcriptase reaction it added, with exception of reverse transcriptase enzyme itself. This control will indicate the level of gDNA contamination (Qiagen, PCR protocols & applications). In this tube, since mRNA was not converted into cDNA, as no reverse transcriptase is present, *taq* polymerase used for qPCR will not have template as it does not amplify mRNA. However, if an amplification curve is visible, we can conclude that this is because our sample is contaminated with gDNA. It is very difficult, however, to have no amplification in this control as is virtually impossible to get rid of all contaminating gDNA. Hence, it was defined that, if a signal in the NRT control is detected 10 cycles apart from the specific signal, gDNA contamination is considered irrelevant (Eurogentec, qPCR guide).

Among positive controls, you can find endogenous and exogenous as well as internal controls. Endogenous positive controls are secondary target genes for which expression level is measured in the same sample of the genes of interest. These genes are used to correct for sample-to-sample variations being, therefore important to have a constitutive expression. As you probably already noted, this is the reference gene use for normalization method described earlier. Exogenous positive controls consist in different samples from the ones under study, and where the gene of interest is known to be expressed. It can be a previous PCR product, RNA transcripts or a plasmid containing the gene of interest (Eurogentec, qPCR guide). This control will permit checking the quality of the primers and reactions. Finally, in the case of the internal positive control, also known as spiking control, DNA or RNA is introduced into the sample, during RNA isolation or immediately before qPCR quantification, and amplified using specific primers. Any variation in this control indicates the presence of inhibitors in your samples (Scipioni, 2008).

Detection chemistries systems

As explained in chapter 2, the detection of the products amplified in qPCR is achieved by the binding of a fluorescent molecule. Detections systems can be divided into 2 main groups: specific and non-specific fluorescent technologies. Specific systems, also called probe-based detection systems, only quantify the desired product and neither unspecific products, such as primer dimer nor products resultant from unspecific primer binding, are quantified (Sigma, qPCR Technical guide). Due to their high specificity for the product, probe-based chemistries allow multiplexing, where more than one gene is detected in the same well. The most well-known probe-based systems are TaqMan®, molecular beacons and FRET probes. However, these systems are very expensive are more difficult to design (Sigma, qPCR Technical guide).

Consequently, due to the lower price and easiness to use, as you only need to design a regular pair of primers instead of a probe, non-specific chemistries are widely used. SYBR® Green, a double stranded DNA intercalating dye, is the most used dye of all. It emits higher levels of fluorescence (approximately 1000×) when bound to double stranded molecules (Agilent Technologies, Introduction to qPCR. Methods and application guide). Though, if unspecific products or primer dimers are present, SYBR® Green will bind and bright equally. Nonetheless, if multiplexing is not a pre-requisite, using the right controls, SYBR® Green will work perfectly. However, when using SYBR® Green as detection chemistry, the generation of a melt curve is mandatory (Agilent Technologies, Introduction to qPCR. Methods and application guide). A melt curve is generated by submitting the final qPCR products to a sequential temperature increase ranging from 65 °C to 95 °C. With the increase in temperature, hydrogen bounds will break and the double stranded molecule become denatured ("melted"). This leads to the dye's dissociation and a proportional decrease in fluorescence intensity (Bio-Rad, Real-time PCR. Applications guide). If one single peak is observed it means that all the products in your sample have the same sequence (figure 4.4A). On the other hand, when the reaction is not specific or efficient, more than one product (hopefully with different sizes and sequences) will be produced and more than one peak will be visualized in the melt peak (figure 4.4B) (Agilent Technologies, Introduction to qPCR. Methods and application guide).

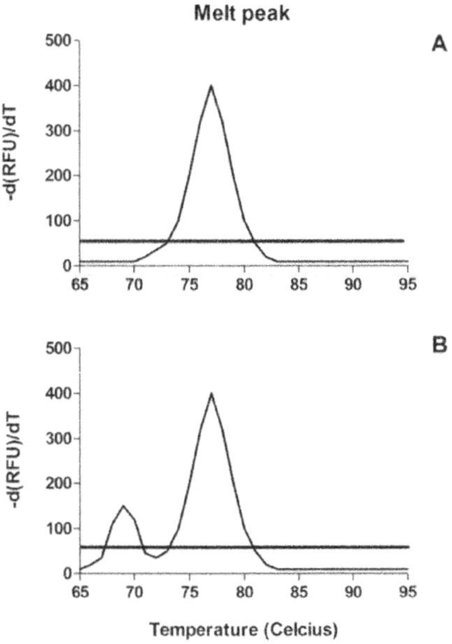

Figure 4.4 | Melt peak of the products amplified in a SYBR Green-based assay. The negative first derivative of the alterations in fluorescence is plotted as a function of the temperature. RFU-relative fluorescence units.

In this situation, several optimizations need to be performed, such as using different melting temperatures or different primer concentrations, or even choosing a new pair of primers.

Amplification efficiency

The reliability of a gene expression quantification assay is directly related with the amplification efficiency of the qPCR run. Therefore, determining the efficiency of your reaction is of the upmost importance. Assuming that the amount of target cDNA doubles after each cycle, it means that the efficiency of amplification is 100%. However, this is not always observed. If amplification efficiency is below 100%, this means that at each cycle the amount of amplification is not doubled (90% efficiency, 1.9 molecules ate each cycle, 85%, 1.85 molecules per each cycle, and so on) (Bio-Rad, Real-time PCR. Applications guide). The main reasons for such inefficient amplification are poor primer design or not-optimal reagent concentration (Bio-Rad, Real-time PCR. Applications guide). There are several methods available to assess amplification efficiency. The most used is the dilution curve method. This method consists in running a set of serial dilutions of a target cDNA with known concentrations, determining their Ct and plotting them on a logarithmic scale. A linear regression is then drawn and using the slope value obtained, amplification efficiency (E) can be determined by applying the following equation:

$$E = 10^{(-1/slope)}$$

If you have prepared 10× serial dilutions, and if the reaction is 100% efficient, the Ct values of each dilution should be 3.3 cycles spaced (2^n= dilution factor; $2^{3.3}$=10) as show in figure 4.5 (Bio-Rad, Real-time PCR. Applications guide).

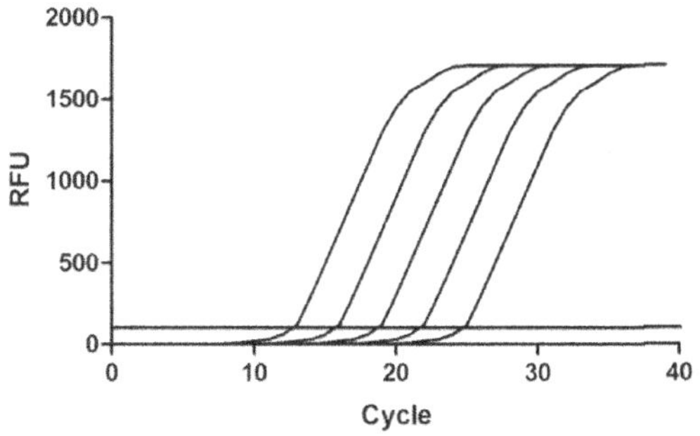

Figure 4.5 | Efficiency determination by the dilution method. Amplification curves of 1:10 cDNA dilutions. Each curve is equally separated by approximately 3.3 cycle indicating 100% efficiency of the reaction.

Even though efficiency close to 100% is an indication of reproducibility and reliability, efficiencies values of about 90% to 110% (corresponding to slopes of 3.1 to 3.6) are considered satisfactory (Life Technologies, Poor PCR efficiency). Lower reaction efficiencies can be due to suboptimal conditions, defective primer design or related with the length and Guanine and Cytosine (GC) contents of the amplicon (Bio-Rad, Real-time PCR. Applications guide). On the other hand, efficiencies higher than 100% may be related with pipetting errors, saturated system, which can be caused by the presence of too much target cDNA in the reaction or by non-specific amplification (Bio-Rad, Real-time PCR. Applications guide). The presence of inhibitors of the polymerase enzyme inhibitors such as ethanol and phenol are another source of efficiencies values higher than 100%.

Primer design

For robust primer design, it is essential to follow several rules. Primers can be designed using different software including free software such as Primer3, Primer-BLAST and Oligo Perfect™ Designer. Independently of the platform used, these are important criteria that should be taken into account (Thornton and Basu, 2011) being the most important listed below:

- Look for secondary structures and avoid design primers on those places. Significant secondary structure may hinder primers from annealing and prevent complete product extension by polymerase. The use of Mfold software can be helpful;
- Primers length should be between 18–28 nucleotides. Shorter primers increase the chance of nonspecific primer binding. Primers too big can hybridize with some mismatching decreasing specificity, and hybridize slower, decreasing yield. Primers need to be long enough for the sequence to be unique and to reduce the probability of the sequence being found at non-target sites;
- Amplicons should present 80-200 bases in length. This allows faster and more efficient reactions;
- The percentage of GC content should be around 40-50%, or mirror the GC content of the template in order to increase binding stability;
- Design primers with Temperature melting (TM) between 59-68°C;
- Choose forward and reverse primers with Tm within 1-3°C from each other and never higher than 4 °C;
- Optimal annealing temperatures are around 60°C;

- Look for primers with GC clamps at the 5' end. Having 1 or 2 GC clamps, in maximum, is ideal since it gives stability to primer binding process;
- Dinucleotide repeats (AGAGAGAGAG) and runs (TCCCCCAG) should be avoided because they promote mispriming. Avoid to have more than 3 to 4 dinucleotides together;
- Avoid mismatches at the 3′ end. The last 3′ nucleotide needs to anneal to the template for optimal extension;
- Check for sequence complementarity bgetween primers and 3' self-complementarity. Primer dimer formation as well as secondary structures should be avoided. Primer dimers formation increase fluorescence in SYBR Green-based runs, resulting in decreased reaction efficiency and inaccurate quantification;
- Check primers specificity using *n*BLAST search.

Experimental protocol

After reading all about qPCR, it's time to prepare the run. In this particular optimized protocol, we used a reference gene normalization approach and SYBR® Green dye for amplification quantification. Please follow the instructions indicated below.

- Dilute your cDNA as well as NRT controls at least 1:100, in ultrapure water;

! be aware ! Too concentrated sample will inhibit qPCR reaction and result in non-reproducible results. The optimal cDNA dilution needs to be experimentally determined (Franca, 2012b).

- Prepare SYBR Green master mix for each of the genes under study by mixing the following components:

2 µL H_2O × *n*
0.5 µL forward primer (10 µM) × *n*
0.5 µL reverse primer (10 µM) × *n*
5 µL of SYBR® Green × *n*

! be aware ! Do not forget to count with NRT (for each different sample) and NTC controls (for each different gene of interest). As before, we advise you to calculate the volumes for n+1 reactions.
! be aware ! Always make your calculations for 1 additional tube. If the number of samples is higher than 20, add 2 more tubes to the total number of tubes needed.

- Define carefully the plate scheme in your laboratory book;
- Prepare the protocol on the qPCR software;
- Define run settings:

94°C during 5 minutes
39 cycles of:
94°C during 15 seconds
60°C during 20 seconds
72°C during 25 seconds

- Define melt curve settings:
  65°C to 95°C
  Increment of 1°C
  Plate reading for 5 seconds
- Pipette, accurately, 2 µL of 1:100 diluted cDNA or NRT control to each appropriated well. Each condition, including controls, needs to be evaluated, at least, in

duplicate to determine reproducibility. Ideally, avoid replicates that have more than 0.4 Ct of difference;

- Add 8 μL of each master mix into the appropriated wells. Avoid bubbles. This may interfere with reading if they persist during the protocol. If possible, perform 1 minute centrifugation to your plate or tubes.
- Cover the plate with the seal plastic;
- Run the protocol.

For more information regarding qPCR guidelines, please read the minimum information for publication of qPCR experiments (MIQE) (Bustin, 2009). This document is promoting experimental transparency, consistency between the results obtained in different research groups and the integrity of the scientific literature.

Data Analysis

At the end of qPCR run, all controls need to be evaluated and you need to confirm if all negative and positive controls are acceptable. Furthermore, you need to observe the melting curve profile to evaluate reaction specificity. After confirming that all controls meet qPCR criteria, you can safely proceed to data analysis.

There are several mathematical models that can be used for relative gene expression quantification. Livak method (Livak and Schmittgen, 2001), also known as $2^{-\Delta\Delta Ct}$, is the most used model. However, this mathematical model assumes that amplification efficiencies, of both reference and target genes, are close to 100% and do not differ from each other more than 5-10% (Livak and Schmittgen, 2001)

If you want to use this mathematical model, you first need to ensure the assumptions postulated are met. Hence, amplification efficiencies of both reference and the target genes need to be experimentally determined before. If the initial assumptions are observed, then you can safely use this formula to quantify gene expression:

$$\text{Relative expression ratio (R)} = 2^{-\Delta\Delta CT}$$

$$\text{Where, } \Delta\Delta Ct = (Ct_{target} - Ct_{reference})_{control} - (Ct_{target} - Ct_{reference})_{test}$$

As an alternative, there is a variation of $2^{-\Delta\Delta Ct}$ method, the comparative $2^{\Delta Ct}$ method that basically yields the same results but is simpler to apply. The mathematical postulations of this method are the same to those of Livak method (Bio-Rad, Real-time PCR. Applications guide).

$$\text{Normalized expression} = 2^{Ct\,(reference) - Ct\,(target)}$$

$$Ratio = \frac{\text{normalized expression test sample}}{\text{normalized expression control sample}}$$

Nevertheless, if both reference and target genes have significant different amplification efficiencies (more than 10%), the methods described above cannot be applied. In such cases, the reaction efficiency (E) is determined experimentally for each reaction (Bio-Rad, Real-time PCR. Applications guide).

$$R = \frac{(\text{E target})^{\Delta Ct\ target\ (Control-sample)}}{(\text{E reference})^{\Delta Ct\ ref\ (Control-sample)}}$$

If there is the need to analyze multiple samples, Pfaffl (Pfaffl, 2004) formulae must be used:

$$R = \frac{(\text{E target})^{\Delta \text{Ct target (MEAN Control- MEAN sample)}}}{(\text{E reference})^{\Delta \text{Ct ref (MEAN Control- MEAN sample)}}}$$

Furthermore, if using more than one reference gene the follwing formula has to be applied (Pfaffl, 2004):

$$R = \frac{(\text{E target})^{\Delta \text{Ct target (MEAN Control- MEAN sample)}}}{(\text{E reference index})^{\Delta \text{Ct ref index (MEAN Control- MEAN sample)}}}$$

Conclusion

Hopefully, as you now understand, by reading this chapter, there are several optimizations and validations that you need to perform before actually starting your experiments. This can take some time and economic resources but is a fundamental stage to obtain reliable and biologically significant results. In complementation of the chapter 3, we hope you now have the right idea of how to optimize your gene expression studies.

Acknowledgments

AF holds a fellowship from the research project EXPL/BIA-MIC/0101/2013 (FCOMP-01-0124-FEDER-041246).

References

Agilent Technologies. Introduction to qPCR. Methods and applications guide.

http://www.chem.agilent.com/Library/brochures/Brochure_Guide%20to%20QPCR_IN70200C.pdf (last accessed in November 2014)

Battaglia M, Pedrazzoli P *et al.* (1998) "Epithelial tumour cell detection and the unsolved problems of nested RT-PCR: a new sensitive one step method without false positive results" in **Bone Marrow Transplant**, 22:693-698

Bhagwat AA, Ying ZI *et al.* (2013) "Determining RNA quality for NextGen sequencing: some exceptions to the gold standard rule of 23S to 16S rRNA ratio" in **Microbiology Discovery**, 1:10

Bioline. Working with RNA: hints and tips. http://www.bioline.com/sg/rna-hints-and-tips (last accessed in November 2014)

Bio-Rad. Real-time PCR applications guide. Bulletin 5279. http://www.biorad.com/webroot/web/pdf/lsr/literature/Bulletin_5279.pdf (last accessed in November 2014).

Bustin SA (2000) "Absolute quantification of mRNA using real-time reverse transcription polymerase chain reaction assays" in **J Mol Endocrinol**, 25:169-193

Bustin SA, Benes V *et al.* (2009) "The MIQE guidelines: minimum information for publication of quantitative real-time PCR experiments" in **Clin Chem**, 55:611-622

Bustin SA, Benes V *et al.* (2005) "Quantitative real-time RT-PCR--a perspective" in **J Mol Endocrinol**, 34:597-601

Carvalhais V, Delgado-Rastrollo M *et al.* (2013) "Controlled RNA contamination and degradation and its impact on qPCR gene expression in *S. epidermidis* biofilms" in **J Microbiol Methods**, 95:195-200

Cheng ZF and Deutscher MP (2005) "An important role for RNase R in mRNA decay" in **Mol Cell**, 17:313-318

Eurogentec. qPCR guide. http://www.eurogentec.com/uploads/qpcr-guide.pdf (last accessed November 2014).

Fleige S and Pfaffl MW (2006) "RNA integrity and the effect on the real-time qRT-PCR performance" in **Mol Aspects Med**, 27:126-139

Franca A, Bento J *et al.* (2012a) "Variability of RNA quality extracted from biofilms of foodborne pathogens using different kits impacts mRNA quantification by qPCR" in **Current Microbiology**, 65:54-59

Franca A, Freitas AI *et al.* (2012b) "Optimizing a qPCR gene expression quantification assay for S. epidermidis biofilms: a comparison between commercial kits and a customized protocol" in **PLoS One**, 7:e37480-

Franca A, Melo L *et al.* (2011) "Comparison of RNA extraction methods from biofilm samples of *Staphylococcus epidermidis*" in **BMC Research Notes**, 4:572-

Heid CA, Stevens J *et al.* (1996) "Real time quantitative PCR" in **Genome Res**, 6:986-994

Huggett J, Dheda K *et al.* (2005) "Real-time RT-PCR normalisation; strategies and considerations" in **Genes Immun**, 6:279-284

Jahn CE, Charkowski AO *et al.* (2008) "Evaluation of isolation methods and RNA integrity for bacterial RNA quantitation" in **J Microbiol Methods**, 75:318-324

Jones LJ, Yue ST *et al.* (1998) "RNA quantitation by fluorescence-based solution assay: RiboGreen reagent characterization" in **Anal Biochem**, 265:368-374

Life Technologies. The Basics: RNase control. http://www.lifetechnologies.com/pt/en/home/references/ambion-tech-support/nuclease-enzymes/general-articles/the-basics-rnase-control.htm (last accessed in November 2014)

Life Technologies. Top 10 ways to improve your RNA isolation. http://www.lifetechnologies.com/pt/en/home/references/ambion-tech-support/rna-isolation/general-articles/ten-ways-to-improve-your-rna-isolation.html (last accessed in November 2014)

Life Technologies. Nucleic acid quantification in solution.

http://www.lifetechnologies.com/content/dam/LifeTech/global/technical-reference-library/Molecular%20Probes%20Handbook/chapter-pdfs/Ch-8-Nucleic-Acid-Detection-Analysis.pdf?icid=WE216841 (last accessed in November 2014)

Life Technologies. RNA priming strategies. http://www.lifetechnologies.com/pt/en/home/life-science/pcr/reverse-transcription/rna-priming-strategies.html (last accessed in November 2014)

Life Technologies. Now it's easy to make your RNA free of genomic DNA contamination and ready for RT-PCR. http://www.lifetechnologies.com/pt/en/home/references/ambion-tech-support/rna-isolation/tech-notes/a-new-method-to-remove-dna.html (last accessed in November 2014)

Life Technologies. Poor PCR efficiency. http://www.lifetechnologies.com/pt/en/home/life-science/pcr/real-time-pcr/qpcr-education/real-time-pcr-troubleshooting-tool/gene-expression-quantitation-troubleshooting/poor-pcr-efficiency.html (last accessed in November 2014)

Livak KJ and Schmittgen TD (2001) "Analysis of relative gene expression data using real-time quantitative PCR and the 2(-Delta Delta C(T)) Method" in **Methods**, 25:402-408

Matos RG, Barria C *et al.* (2014) "The importance of proteins of the RNase II/RNB-family in pathogenic bacteria" in **Front Cell Infect Microbiol**, 4:68-

Pfaffl MW (2004) "Quantification strategies in real-time PCR" p. 87-112. In SA Bustin (Eds.), A-Z of quantitative PCR, International University Line, La Jolla, CA, USA.

Pfaffl MW (2014) "Relative quantification" p. 63-82. In T Dorak (Eds.), Real-time PCR, International University Line, La Jolla, CA, USA.

Pinto FL, Thapper A *et al.* (2009) "Analysis of current and alternative phenol based RNA extraction methodologies for cyanobacteria" in **BMC Mol Biol**, 10:79

Promega. Calculating nucleic acid or protein concentration. https://worldwide.promega.com/resources/pubhub/applications-notes/calculating-nucleic-acid-or-protein-concentration-using-the-glomax-multi-plus-microplate-instrument/(last accessed in November 2014)

Qiagen. RNeasy mini kit handbook 4th edition. http://www.genome.duke.edu/cores/microarray/services/rna-qc/documents/RNeasy_Mini_Handbook.pdf

Qiagen. PCR protocols & applications. http://www.qiagen.com /pt/resources/molecular-biology-methods/pcr/ (last accessed in November 2014)

Rump LV, Asamoah B *et al.* (2010) "Comparison of commercial RNA extraction kits for preparation of DNA-free total RNA from *Salmonella* cells" in **BMC Res Notes**, 3:211

Schoor O, Weinschenk T *et al.* (2003) "Moderate degradation does not preclude microarray analysis of small amounts of RNA" in **Biotechniques**, 35:1192-1201

Schroeder A, Mueller O *et al.* (2006) "The RIN: an RNA integrity number for assigning integrity values to RNA measurements" in **BMC Mol Biol**, 7:3

Scipioni A, Mauroy A *et al.* (2008) "A SYBR Green RT-PCR assay in single tube to detect human and bovine noroviruses and control for inhibition" in **Virol J**, 5:94

Sieber MW, Recknagel P *et al.* (2010) "Substantial performance discrepancies among commercially available kits for reverse transcription quantitative polymerase chain reaction: a systematic comparative investigator-driven approach" in **Anal Biochem**, 401:303-311

Sigma, qPCR Technical guide. http://www.sigmaaldrich.com/content/dam/sigma-aldrich/docs/Sigma/General_Information/qpcr_technical_guide.pdf (last accessed in November 2014)

Souaze F, Ntodou-Thome A *et al.* (1996) "Quantitative RT-PCR: limits and accuracy" in **Biotechniques**, 21:280-285

Thellin O, Zorzi W *et al.* (1999) "Housekeeping genes as internal standards: use and limits" in **J Biotechnol**, 75:291-295

Thermo Scientific. Basic principles of RT-qPCR.
http://www.thermoscientificbio.com/applications/basic-rt-qpcr/ (last accessed in November 2014)

Thornton B and Basu C (2011) "Real-time PCR (qPCR) primer design using free online software" in **Biochem Mol Biol Educ**, 39:145-154

Vandesompele J, De Paepe A *et al.* (2002a) "Elimination of primer-dimer artifacts and genomic coamplification using a two-step SYBR green I real-time RT-PCR" in **Anal Biochem**, 303:95-98

Vandesompele J, De PK *et al.* (2002b) "Accurate normalization of real-time quantitative RT-PCR data by geometric averaging of multiple internal control genes" in **Genome Biol**, 3:RESEARCH0034

Wacker MJ and Godard MP (2005) "Analysis of one-step and two-step real-time RT-PCR using SuperScript III" in **J Biomol Tech**, 16:266-271

Wilfinger WW, Mackey K *et al.* (1997) "Effect of pH and ionic strength on the spectrophotometric assessment of nucleic acid purity" in **Biotechniques**, 22:474-481

Wong ML and Medrano JF (2005) "Real-time PCR for mRNA quantitation" in **Biotechniques**, 39:75-85

Zhang J and Byrne CD (1999) "Differential priming of RNA templates during cDNA synthesis markedly affects both accuracy and reproducibility of quantitative competitive reverse-transcriptase PCR" in **Biochem J**, 337 (Pt 2):231-241

Transcriptomes of relevant medical biofilms

Section II

Chapter 5

Introduction to medical biofilms

Nuno Cerca
CEB - Centre of Biological Engineering, LIBRO - Laboratory of Research in Biofilms Rosário Oliveira, University of Minho, Campus de Gualtar, 4710-057 Braga, Portugal
nunocerca@ceb.uminho.pt

Contents of chapter

The origin of bacterial biofilms

Most certainly bacterial biofilms exists since the dawn on life but our awareness of these bacterial communities is, on a cosmic scale, pretty recent. Although the presence of microscopic life was long suspected, only after the invention of the microscope, in the XVII century, the confirmation of microscopic single cell organisms was possible. However, the origins of these single cell organisms were still elusive and only in the XIX century it became clear the significance of microbial life. To this achievement, credits go to French and German scientists Louis Pasteur and Robert Koch.

Pasteur introduced the concept of sterilization and was able to disprove the theory of spontaneous generation that defended that microorganisms were created from non-living materials when the conditions were appropriate (Pasteur, 1881). His findings were on the origin of attributing life as we know it to microorganisms. A few years later Koch introduced the concept of a pure culture. When observing single colonies in solid media, he proposed that each colony derived from a single microorganism. By then, it was already thought that some diseases were caused by microorganisms. When studying anthrax, a disease commonly associated with cattle, he found out that by injecting a health cow with blood from a sick cow, the healthy cow would soon become sick (Koch, 1884). By observing that the sick cow's blood contained live microorganisms, he came up with a series of postulates that are nowadays known as the Koch's postulates. These were:

1. The suspected pathogenic organism should be present in all cases of the disease and absent from healthy animals;
2. The suspected organism should be grown in pure culture;
3. Cells from a pure culture of the suspected organism should cause disease in a healthy animal;
4. The organism used to cause disease should be reisolated from the animal and shown to be the same as the original.

Koch's view of the infectious disease process shaped the clinical microbiology science for over a century. In fact, actual microbiology culture techniques are still based on the pure culture grown in solid media Petri dishes.

It was only in the late 1970s that environmental microbiologists observed the presence of adhered microbial communities in many natural habitats. With the development of more advanced analytical techniques, such as confocal laser scanning microscopy, new insights on biofilm physiology and structure were achieved. At that time it was surprisingly to see that biofilms communities where found to have complex heterogeneous tridimensional structures (reviewed in Costerton, 1995).

These findings lead the way to the beginning of biofilm science. A few years after the description of biofilm communities in the environmental microbiology field, the first reports of microbial biofilms associated with disease came to light, and currently, pathogenic bacterial biofilms are acknowledged to have an important role on infection, as will be further discussed in this book.

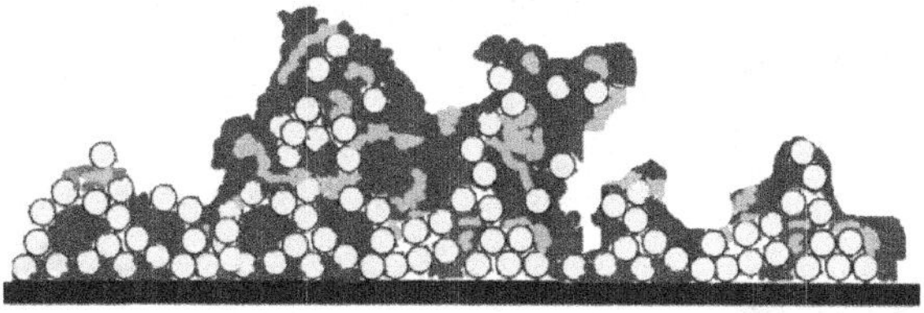

Figure 5.1 | Representation of the structure of a bacterial biofilm. Bacterial cells are shown as white circles and the dark and light gray represent the matrix components.

How a biofilm is formed

A bacterial biofilm can be defined as a community of bacteria lived attached to a surface and surrounded by a matrix of extrapolimeric substances, such as proteins, polysaccharides and nucleic acids. These substances can either be secreted by the biofilm bacteria or entrapped in the matrix during bacterial growth (reviwed in Hall-Stoodley, 2004). When planktonic bacteria come in contact with a surface, a certain amount of cells will be attached to the surface. This initial attachment is often mediated by non-specific interactions, namely hydrophobic and electrostatic interactions, among others (reviwed in Bos, 1999). The nature of this non-specific interactions results in the fact that virtually any bacteria can adhere (to some extent) to a surface. This is the first stage of biofilm formation (figure 5.2). For adhered bacteria to further grown in clusters, the presence of specific molecules, generally called intercellular adhesins, is required (Heilmann, 1996).

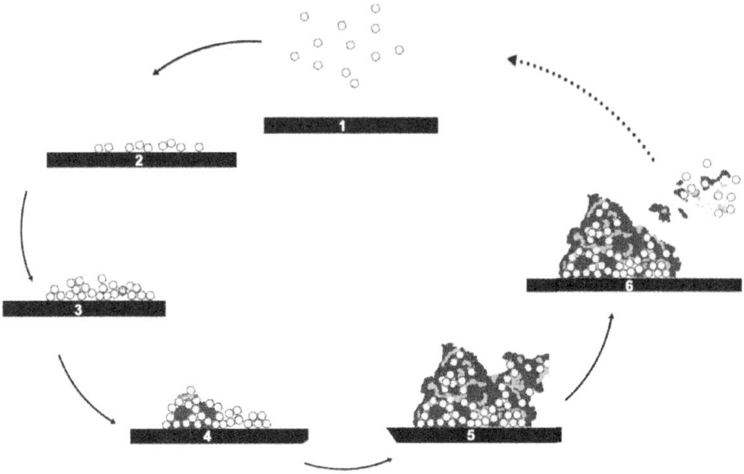

Figure 5.2 | Stages of biofilm formation. Planktonic bacteria (1) will first adhere to a surface (2) followed by growth in clusters (3). With bacterial growth adhered to the surface, the matrix starts to be formed (4). Over time the biofilm becomes thicker and the matrix more prominent (5). Bacteria will then detach from the biofilm and will colonize other surfaces (6).

As bacteria starts to grow adhered to the surface, and then to each other, biofilms starts to mature. Biofilm maturation will involve the formation of the biofilm matrix that will be further discussed in this chapter. Finally, either by the sheer forces of fluid or by specific gene expression, biofilms will release bacteria that will act as new colonizers of different surfaces or regions of a specific surface (reviewed in Abee, 2010).

Biofilm formation is highly regulated and there have been hundreds of genes described to have roles in biofilm formation. Table 5.1 shows a brief example of some genes involved in biofilm formation of pathogenic bacteria.

Table 5.1 | Examples of genes involved in the different stages of biofilm formation.

Bacteria	Gene	Role in biofilms
Pseudomonas aeruginosa	*psl*	Biofilm maturation
	lasI	Quorum sensing
Escherichia coli	*wcaB*	Biofilm structure
	agn43	Aggregation
Staphylococcus epidermidis	*icaADBC*	Aggregation
Staphylococcus aureus	*sasG*	Initial adhesion
Streptococcus gordonii	*abpA*	Initial adhesion
Streptococcus mutans	*gtfB*	Aggregation
	luxS	Quorum sensing

Why pathogenic bacteria form biofilms?

Laboratory experiments have determined that biofilm formation is often an energetic demanding process (reviewed in Bowden and Li, 1997). For instance, most clinical isolates of *Staphylococus aureus* only form biofilms in the presence of extra glucose (Croes, 2009). Having this in consideration, it seems obvious that, on an evolutionary point of view, biofilms have to provide several advantages to bacteria, otherwise individual bacteria would probably not join the community.

One of the most well-known described advantages for biofilm bacteria has to do with the matrix. The biofilm tridimensional structure normally acts as a physical barrier against external stimuli (reviewed in Flemming, 2007). Bacteria embedded in the matrix will be most likely to withstand the sheer forces of the hydrodynamic forces of the physiological fluids such as blood or saliva. Maintaining adhesiveness to a surface in a permissive environment seems to be an evolutionary advantage for bacteria.

Furthermore, the matrix has been associated with the ability of bacteria to survive antibiotic treatment (reviewed in Gilbert, 1997) and also the host immune response (Cerca, 2006). This is, in fact, one of the main reasons so many opportunistic pathogens are now considered important human pathogens when associated with biofilm infections, such as *S. epidermidis*.

More than providing with a physical protection, the biofilm matrix also provides advantages to bacteria when it comes to nutritional requirements. Due to the tridimensional structure of biofilms, water and other nutrients reservoirs are easily formed inside the biofilms, providing a stable energy source to bacteria (reviewed in Flemming, 2007).

Biofilms have also been shown to be an excellent source of external DNA acquisition (Steinberger and Holden, 2005). Such as all other superior live forms, bacteria have evolutionary advantages by acquiring new genetic traits. One such example, where a mixed species bacterial biofilms communities have demonstrated gene transfer, has been the oral plaque, one of the most well know pathogenic biofilms (Zijnge, 2010).

The biofilm matrix

As described in the previous sections, the biofilm matrix arises on the final stages of biofilm formation. Since the first observation of bacterial biofilms that the matrix has been noted as crucial for the survival of the bacteria inside the biofilm, against adverse conditions, such as low nutrient availability, high osmotic stress or even in the presence of antibiotics, as further discussed in this chapter.

Table 5.2 | Main functions of the biofilm matrix components

Function	Matrix components
Adhesion	Proteins, DNA and polysaccharides
Protective barrier	Proteins and polysaccharides
Water retention	Polysaccharides
Enzymatic activity	Proteins
Nutrient source	Proteins, DNA, polysaccharides, and lipids
Energy stockpiling	Polysaccharides

In fact, the importance of the biofilm matrix can be elucidated by the estimated 90% of dry-weight mass that is attributed to the matrix itself. The matrix composition is highly variable, since it depends both on the environment where the biofilm is been formed and also on the specific bacterial strains involved in the process. Nevertheless, most biofilm matrices are composed by extracellular proteins, polysaccharides and DNA (Sutherland, 2001).

Individual components of biofilm matrices have been linked to specific functions (table 5.2). However, not all biofilm matrices components have been proper studied and in fact, little is known regarding molecular mechanisms of matrix formation.

The quorum sensing phenomena

In most Gram-negative bacteria, biofilm formation is controlled by a central mechanism known as quorum sensing (Jayaraman and Wood, 2008). Quorum sensing literally means recognition of nearby cell density. As noted above, biofilm formation has a high energetic consumption and before bacteria attempts to form a biofilm, all the proper conditions should be present. When bacteria are present in high density, such as biofilms, quorum sensing allows the individual bacteria to shift its transcriptome in order to favor the expression of important genes for the biofilm mode of growth (Wagner, 2003).

Bacteria with quorum sensing abilities produce extracellular peptides known as autoinducers, usually produced at low levels. However, when a bacterial population reaches high densities, the concentration of the autoinducers obviously also reach higher titers. Under these conditions, the chance of an autoinducer to be recognized by specific cell surface associated proteins will be higher. When this occurs, the autoinducer will be transported into the cell cytoplasm and will activate specific mRNA transcription, including its own expression. In this way, more autoinducer peptides will be produced, increasing the possibility of inducing other cells to undergo the same process, further stimulating biofilm formation(Nadell, 2008).

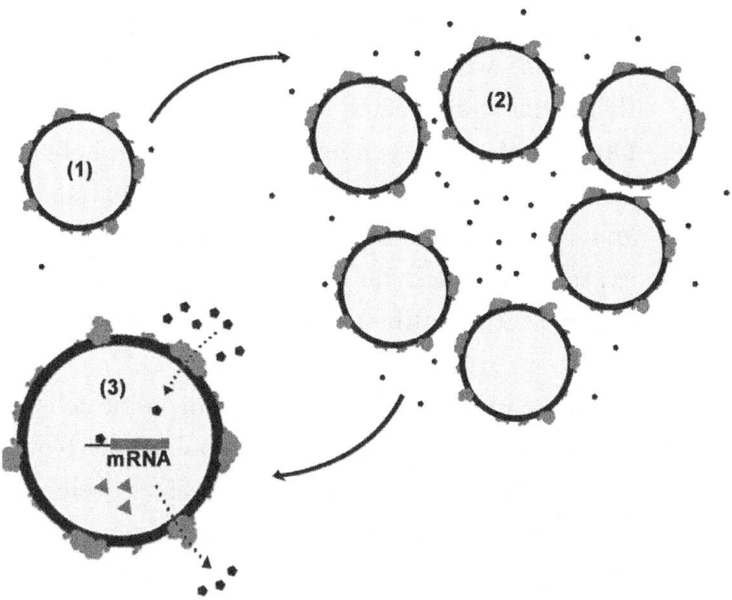

Figure 5.3 | The quorum sensing phenomena. Bacteria produce recognition signals at low densities (1); when a bacteria population reaches a higher density, the recognition signal will also reach a critical level (2); the recognition signal at high densities will enter the cell cytoplasm and will induce the transcription of specific mRNA related with biofilm formation (3). This will further induct the production of the recognition signal, boosting up the biofilm forming cycle.

The impact of biofilms in health

As described above, biofilm bacteria are normally resistant to antibiotics. Contrary to traditional genetic resistance mechanisms, biofilm resistance to antibiotics seems to be related with the tridimensional structure of the biofilm and is multi-factorial, including the following aspects:

1. Impaired diffusion of antibiotics trough the matrix (Zheng and Stewart, 2002);

2. Heterogeneous bacterial growth implying that some cells will be metabolically inactive, rendering them almost unaffected by antibiotics (Cerca, 2014);
3. Due to the microenvironment provided by the biofilm, some cells will be under external stresses and will constantly induce differential gene expression, such as efflux pumps (that will transport the antibiotics out of the cell) (Fonseca, 2014);
4. After the initial antibiotic treatment, some cells will survive and eventually will repopulate the biofilm. These cells are known as persister cells and constitute the base for recurrence on biofilm infections (reviewed in Keren, 2004).

In the healthcare industry, biofilms are often associated with infections with indwelling medical devices, such as central venous catheters or prosthesis (table 5.3). The insertion of medical devices in a patient provides a surface for bacterial colonization and often it's the process of insertion that is responsible for the initial colonization (reviewed in Otto, 2009). Even when prevention measures are taken, the constant contact between the infection site and the outside increases the chances of biofilm development. Since biofilm infections normally only occur in the hospital environment, these infections are also known as nosocomial infections. Nevertheless, nosocomial infections can also include other infections that are not related with biofilms (Goossens, 2005).

Table 5.3 | Examples of some known medical devices related biofilm infections

Bacteria	Pathology	Medical devices
Pseudomonas aeruginosa	*Cystic fibrosis*	Ventilators
Escherichia coli	*Urinary tract infections*	Catheters
Staphylococcus epidermidis	*Systemic infections*	Catheters, prosthesis
Staphylococcus aureus	*Endocarditis*	Heart valves

While biofilms are often associated with medical devices related infections (such as described in chapter 6 or 8), there are some other examples where biofilms are not associated with any medical device, such in the case of bacterial vaginosis, as will be discussed in chapter 11.

Despite concerted efforts to treat biofilm related infections with antibiotic therapy, due to the above mentioned characteristics of biofilms, in most cases the only possible solution is the physical removal of the infected medical device. While this is not commonly life threatening, clearly it is an economic burden to public health systems and also to the patient (Rogers, 2009). In the next chapters some of the most well know biofilm related infections will be discussed in detail.

References

Abee T, Kovacs AT *et al.* (2010) "Biofilm formation and dispersal in Gram-positive bacteria" in **Current Opinion in Biotechnology**, 22:172-9

Bos R, van der Mei H *et al.* (1999) "Physico-chemistry of initial microbial adhesive interactions: its mechanisms and methods for study" in **FEMS Microbiology Reviews**, 23:179-230

Bowden GH and Li YH (1997) "Nutritional influences on biofilm development" in **Advances in Dental Research**, 11:81-99

Cerca F, Franca A *et al.* (2014) "Dormant bacteria within *Staphylococcus epidermidis* biofilms have low inflammatory properties and maintain tolerance to vancomycin and penicillin after entering planktonic growth" in **Journal of Medical Microbiology**, 63:1274-1283

Cerca N, Jefferson K *et al.* (2006) "Comparative antibody-mediated phagocytosis of *Staphylococcus epidermidis* cells grown in a biofilm or in the planktonic state" in **Infection and Immunity**, 74:4849-4855

Costerton J, Lewandowski Z *et al.* (1995) "Microbial biofilms" in **Annual Reviews in Microbiology**, 49:711-745

Croes S, Deurenberg RH *et al.* (2009) "*Staphylococcus aureus* biofilm formation at the physiologic glucose concentration depends on the S. aureus lineage" in **BMC Microbiology**, 9:229-

Flemming HC, Neu TR *et al.* (2007) "The EPS matrix: the "house of biofilm cells"" in **Journal of Bacteriology**, 189:7945-7947

Fonseca E, Silva S *et al.* (2014) "Effects of fluconazole on *Candida glabrata* biofilms and its relationship with ABC transporter gene expression" in **Biofouling**, 30:447-457

Gilbert P, Das J *et al.* (1997) "Biofilm susceptibility to antimicrobials" in **Advances in Dental Research**, 11:160-167

Goossens H (2005) "European status of resistance in nosocomial infections" in **Chemotherapy**, 51:177-181

Hall-Stoodley L, Costerton JW *et al.* (2004) "Bacterial biofilms: from the natural environment to infectious diseases" in **Nature Reviews Microbiology**, 2:95-108

Heilmann C, Schweitzer O *et al.* (1996) "Molecular basis of intercellular adhesion in the biofilm-forming *Staphylococcus epidermidis*" in **Molecular Microbiology**, 20:1083-1091

Jayaraman A and Wood TK (2008) "Bacterial quorum sensing: signals, circuits, and implications for biofilms and disease" in **Annual Review of Biomedical Engineering**, 10:145-167

Keren I, Kaldalu N *et al.* (2004) "Persister cells and tolerance to antimicrobials" in **FEMS Microbiology Letters**, 230:13-18

Koch R (1884) "An Address on Cholera and its Bacillus" in **British Medical Journal**, 2:403-407

Nadell CD, Xavier JB *et al.* (2008) "The Evolution of Quorum Sensing in Bacterial Biofilms" in **PLoS Biology**, 6:171-179

Otto M (2009) "*Staphylococcus epidermidis*--the 'accidental' pathogen" in **Nature Reviews Microbiology**, 7:555-567

Pasteur L (1881) "On the Germ theory" in **Science**, 2:420-422

Rogers K, Fey P *et al.* (2009) "Coagulase-Negative Staphylococcal Infections" in **Infectious Disease Clinics of North America**, 23:73-98

Steinberger RE and Holden PA (2005) "Extracellular DNA in single- and multiple-species unsaturated biofilms" in **Applied and Environmental Microbiology**, 71:5404-5410

Sutherland IW (2001) "The biofilm matrix--an immobilized but dynamic microbial environment" in **Trends Microbiol**, 9:222-227

Wagner VE, Bushnell D *et al.* (2003) "Microarray analysis of *Pseudomonas aeruginosa* quorum-sensing regulons: effects of growth phase and environment" in **Journal of Bacteriology**, 185:2080-2095

Zheng Z and Stewart PS (2002) "Penetration of rifampin through *Staphylococcus epidermidis* biofilms" in **Antimicrobial Agents and Chemotherapy**, 46:900-903

Zijnge V, van Leeuwen MB *et al.* (2010) "Oral biofilm architecture on natural teeth" in **PLoS One**, 5:e9321

Chapter 6

Staphylococcus epidermidis biofilm infections

Virgínia Carvalhais[1,2], Ângela França[1*]
[1]CEB - Centre of Biological Engineering, LIBRO - Laboratory of Research in Biofilms Rosário Oliveira, University of Minho, Campus de Gualtar, 4710-057 Braga, Portugal
[2]Instituto de Ciências Biomédicas de Abel Salazar, Universidade do Porto, Porto, Portugal
*** afranca@deb.uminho.pt**

Contents of chapter

Introduction to the chapter

Nowadays, *Staphylococcus* species are one of the most important causative agents of healthcare-associated infections. Staphylococci are classically divided into coagulase-positive, to which *Staphylococcus aureus* belongs and, coagulase-negative (CoNS), in which *Staphylococcus epidermidis* is the major representative (Kleeman, 1993; Kloos and Bannerman, 1994). *S. epidermidis* colonizes skin and mucosae as a commensal microbe (Grice, 2009; Kloos and Musselwhite, 1975). However, it is commonly isolated in indwelling medical devices-related infections (Rupp and Archer, 1994; Uckay, 2009). This association has to do with *S. epidermidis* ability to adhere and to form highly structured biofilms on the surface of indwelling medical devices (reviewed in Vadyvaloo and Otto, 2005). The self-produced extracellular matrix, where biofilm cells are embedded, represents the main virulence factor as it increases *S. epidermidis* biofilms tolerance to several antibiotics and host immune response (Vuong, 2004). Consequently, these biofilms are often associated with the emergence of chronic and relapsing infections (Cheung, 2010; reviewed in Costerton, 1999), which are extremely difficult to treat (reviewed in Costerton, 1999). It is estimated that about 80% of the chronic bacterial infections diagnosed are biofilm-associated (Davies, 2003). Among them, CoNS are the leading cause of indwelling medical devices-related infections, wherein 80% of cases are attributed to *S. epidermidis* biofilms (Gotz, 2002).

CoNS biofilms impact in clinical setting

Despite the great contribution of indwelling medical devices to the advance and improvement of medicine and healthcare, medical devices represent important focus of infections (Donlan, 2001). *S. epidermidis* biofilm infections are particularly associated with intravenous catheters but also with prosthetic heart valves, joint prostheses, peritoneal dialysis catheters, cardiac pacemakers, cerebrospinal fluid shunts, endotracheal tubes, urinary tract catheters, intraocular lens and other polymer or metal devices (reviewed in Rupp, 2014). Critically ill patients, such as immunocompromised and long-term hospitalized patients, are the most affected. About 95% of urinary tract infections are catheter related, 87% of the cases of bloodstream infections are originated from vascular catheters, and 86% of the cases of pneumonia are associated with mechanical ventilation (Richards, 1999). In Portugal, a 2012 inquiry reported significant differences in the number of infections among patients with, or without, indwelling-medical devices (Pina, 2013). It was found that while 31.3% of the patients with vascular central catheters developed infection, only in 9.7% of the patients without catheter was detected infection. The same tendency was observed for urinary tract catheters, where 32.9% of patients with the dispositive developed infection, compared to 9.7% of the non-catheterized patients.

To worsen the high incidence and prevalence of biofilm-associated infections, therapeutics based on antimicrobials are usually unsuccessful in the eradication of biofilms (reviewed in Costerton, 1999). It is known that *S.*

epidermidis biofilms present higher capacity to tolerate antimicrobial agents than their planktonic counterparts (Cerca, 2005). Several mechanisms were proposed for the increased tolerance to antimicrobial therapy, such as the limited penetration and diffusion of antibiotics through the matrix (reviewed in Stewart and Costerton, 2001) and the wide range of cells' physiological states within a biofilm, which includes aerobically and fermentatively growing cells, dormant and dead cells (Rani, 2007).

Dormant cells present lowered metabolic activity and, because of that, biofilms susceptibility to antibiotics may be impaired as most antimicrobials require metabolically active cells to be effective (Anderson and O'Toole, 2008). Thus, removal of the contaminated devices is often the only way to resolve these infections. Aside from the costs associated with the diagnosis and treatment of biofilm-related infections, these have serious implications in patients' quality of life, resulting, occasionally, in mortality among infected patients (reviewed in Otto, 2009). In Europe countries such as UK, France, Italy and Germany, it was estimated that catheter-related bloodstream infections (responsible for 80% of the cases of bacteraemia diagnosed) resulted in more than 1,000 deaths per year and an annual burden of 35 to 164 million Euros (Tacconelli, 2009).

Diseases caused by CoNS biofilms

S. epidermidis is one of the most common causes of biofilm-related infections, being associated to a wide range of medical devices (table 6.1) (reviewed in Rogers, 2009).

Nevertheless, *S. epidermidis* is particularly associated with intravascular catheter-associated infections being estimated that approximately 30 to 40% of the nosocomial bloodstream infections diagnosed are caused by *S. epidermidis* biofilm formation on intravascular catheters (reviewed in Rupp, 2004). In table 6.1 the percentages of biofilm-related infections caused by *S. epidermidis* in different medical devices are indicated (reviewed in Rupp, 2014).

Table 6.1 | Biofilm-associated infections caused by *S. epidermidis*.

Biofilm infection	% of infection
Intravascular catheter infection and bacteraemia	30 – 40%
Endocarditis, cardiac devices infections and vascular graft	15 – 40%
Orthopedic prosthetic devices infections	30 – 43%
Peritoneal dialysis catheter- associated infections	20 – 40%
Genitourinary prostheses (synthetic urinary sphincters and penile prostheses)	35 – 60%
Central nervous system shunt infections	1.7 – 2.5%

S. epidermidis biofilms were also found in surgical site infections, breast implants and mastitis, and endophthalmitis associated with intraocular lens. Furthermore, nenonates may also develop urinary tract infections, wound abscesses, pneumonia, meningitis, enterocolitis and omphalitis mostly caused by *S. epidermidis* biofilms (reviewed in Rogers, 2009 and Rupp, 2014).

Transcriptomics of *Staphylococcus epidermidis* biofilm-related infections

Transcriptomics has significantly contributed to decipher biofilms dynamics and biofilm-related infections particularities. One of the first transcriptomic studies performed in *S. epidermidis* biofilms analyzed the particular phenotypic alterations occurring during the transition from planktonic to biofilm phenotype (Yao, 2005). As could be expected, the down-regulation of transcription, translation and aerobic production of energy was found in *S. epidermidis* biofilms. The low biofilm metabolic activity, due to the higher proportion of dormant cells, was seen as a mechanism of adjustment to external conditions and to promote long-term survival.

Dormancy has also important implications in host immune response. It was found that *S. epidermidis* biofilms with higher proportions of dormant cells induced a lower activation of immune cells, such as macrophages and dendritic cells, showing that dormancy contributes to *S. epidermidis* biofilms evasion of the host immune system (Cerca, 2011; Cerca, 2014). In order to identify the genetic basis of such differences, gene expression profiling of *S. epidermidis* biofilms with higher and lower proportion of dormant cells was assessed by RNA-sequencing (Carvalhais, 2014). Interestingly, it was shown that biofilms with higher proportion of dormant cells presented an increased expression of genes related to oxidation-reduction processes and a decreased expression of genes involved in translation processes (Carvalhais, 2014).

Along with the high physiological heterogeneity of biofilm cells, biofilms characteristic multilayered and dense structure, together with matrix components, also provides protection against host immune response effectors. Indeed, it was shown that *S. epidermidis* biofilms are capable to evade human phagocytes (Cheung, 2010) and opsonins (Kristian, 2008). Several immunological studies have been performed in order to better understand *S. epidermidis* biofilm-related infections pathogenesis. As *S. epidermidis* is a frequent cause of vascular catheter-associated bloodstream infections, the study of the interaction between the bacterium and human blood is essential to specifically target these infections. Recently, using an *ex vivo* human blood model, the interaction between *S. epidermidis* biofilms and human blood circulating immune system effectors was assessed (Franca, 2014). Interestingly, 4 hours after biofilm contact with human blood, it was observed an increased expression of genes involved in iron sequestration and usage. This result suggests that iron is essential for *S. epidermidis* biofilms survival in human blood. Similar conclusions were drawn for other pathogens. Consequently, vaccine strategies targeting molecules capable of sequestering iron from the host are being developed (Alteri, 2009; Stranger-Jones, 2006; Thompson, 2003).

Even though the study of bacteria' phenotype during infection gives invaluable information, to fully understand biofilm-related infections, the characterization of the host immune system response to biofilm infections is also necessary. However, at the best of our knowledge, no studies concerning host gene expression studies during *S. epidermidis* biofilm infections were reported so far.

Conclusion

Although transcriptomics studies have greatly increased the current knowledge on biofilm-related infections, much more studies are needed so more efficient strategies for the diagnosis, treatment and prevention of these infections can be conceived.

Acknowledgments

VC holds an individual FCT fellowship (SFRH/BD/78235/2011). AF holds a fellowship from the research project EXPL/BIA-MIC/0101/2013 (FCOMP-01-0124-FEDER-041246).

References

Alteri CJ, Hagan EC *et al.* (2009) "Mucosal immunization with iron receptor antigens protects against urinary tract infection" in **PLoS Pathog**, 5:e1000586

Anderson GG and O'Toole GA (2008) "Innate and induced resistance mechanisms of bacterial biofilms" in **Curr Top Microbiol Immunol**, 322:85-105

Carvalhais V, Franca A *et al.* (2014) "Dormancy within *Staphylococcus epidermidis* biofilms: a transcriptomic analysis by RNA-seq" in **Appl Microbiol Biotechnol**, 98:2585-2596

Cerca F, Andrade F *et al.* (2011) "*Staphylococcus epidermidis* biofilms with higher proportions of dormant bacteria induce a lower activation of murine macrophages" in **J Med Microbiol**, 60:1717-1724

Cerca F, Franca A *et al.* (2014) "Dormant bacteria within *Staphylococcus epidermidis* biofilms have low inflammatory properties and maintain tolerance to vancomycin and penicillin after entering planktonic growth" in **J Med Microbiol**, 63:1274-83

Cerca N, Martins S *et al.* (2005) "Comparative assessment of antibiotic susceptibility of coagulase-negative staphylococci in biofilm versus planktonic culture as assessed by bacterial enumeration or rapid XTT colorimetry" in **Journal of Antimicrobial Chemotherapy**, 56:331-336

Cheung GY, Rigby K *et al.* (2010) "*Staphylococcus epidermidis* strategies to avoid killing by human neutrophils" in **PLoS Pathog**, 6:e1001133

Costerton JW, Stewart PS *et al.* (1999) "Bacterial biofilms: a common cause of persistent infections" in **Science**, 284:1318-1322

Davies D (2003) "Understanding biofilm resistance to antibacterial agents" in **Nat Rev Drug Discov**, 2:114-122

Donlan RM (2001) "Biofilms and device-associated infections" in **Emerg Infect Dis**, 7:277-281

Franca A, Carvalhais V *et al.* (2014) "Alterations in the *Staphylococcus epidermidis* biofilm transcriptome following interaction with whole human blood" in **Pathog Dis**, 70:444-448

Gotz F (2002) "Staphylococcus and biofilms" in **Molecular Microbiology**, 43:1367-1378

Grice EA, Kong HH *et al.* (2009) "Topographical and temporal diversity of the human skin microbiome" in **Science**, 324:1190-1192

Kleeman KT, Bannerman TL *et al.* (1993) "Species distribution of coagulase-negative staphylococcal isolates at a community hospital and implications for selection of staphylococcal identification procedures" in **J Clin Microbiol**, 31:1318-1321

Kloos WE and Bannerman TL (1994) "Update on clinical significance of coagulase-negative staphylococci" in **Clin Microbiol Rev**, 7:117-140

Kloos WE and Musselwhite MS (1975) "Distribution and persistence of Staphylococcus and Micrococcus species and other aerobic bacteria on human skin" in **Appl Microbiol**, 30:381-385

Kristian SA, Birkenstock TA *et al.* (2008) "Biofilm formation induces C3a release and protects *Staphylococcus epidermidis* from IgG and complement deposition and from neutrophil-dependent killing" in **J Infect Dis**, 197:1028-1035

Otto M (2009) "*Staphylococcus epidermidis*--the 'accidental' pathogen" in **Nat Rev Microbiol**, 7:555-567

Rani SA, Pitts B *et al.* (2007) "Spatial patterns of DNA replication, protein synthesis, and oxygen concentration within bacterial biofilms reveal diverse physiological states" in **J Bacteriol**, 189:4223-4233

Richards MJ, Edwards JR *et al.* (1999) "Nosocomial infections in medical intensive care units in the United States. National Nosocomial Infections Surveillance System" in **Crit Care Med**, 27:887-892

Rogers KL, Fey PD *et al.* (2009) "Coagulase-negative staphylococcal infections" in **Infect Dis Clin North Am**, 23:73-98

Rupp ME (2014) "Clinical characteristics of infections in humans due to *Staphylococcus epidermidis*" in **Methods Mol Biol**, 1106:1-16

Rupp ME and Archer GL (1994) "Coagulase-negative staphylococci: pathogens associated with medical progress" in **Clin Infect Dis**, 19:231-243

Stranger-Jones YK, Bae T *et al.* (2006) "Vaccine assembly from surface proteins of *Staphylococcus aureus*" in **Proc Natl Acad Sci U S A**, 103:16942-16947

Stewart P and Costerton J (2001) "Antibiotic resistance of bacteria in biofilms" in **Lancet**, 358:135-138

Tacconelli E, Smith G *et al.* (2009) "Epidemiology, medical outcomes and costs of catheter-related bloodstream infections in intensive care units of four European countries: literature- and registry-based estimates" in **J Hosp Infect**, 72:97-103

Thompson EA, Feavers IM *et al.* (2003) "Antigenic diversity of meningococcal enterobactin receptor FetA, a vaccine component" in **Microbiology**, 149:1849-1858

Uckay I, Pittet D *et al.* (2009) "Foreign body infections due to *Staphylococcus epidermidis*" in **Ann Med**, 41:109-119

Vadyvaloo V and Otto M (2005) "Molecular genetics of Staphylococcus epidermidis biofilms on indwelling medical devices" in **Int J Artif Organs**, 28:1069-1078

Vuong C, Voyich JM *et al.* (2004) "Polysaccharide intercellular adhesin (PIA) protects *Staphylococcus epidermidis* against major components of the human innate immune system" in **Cellular Microbiology**, 6:269-275

Yao Y, Sturdevant DE *et al.* (2005) "Genomewide analysis of gene expression in *Staphylococcus epidermidis* biofilms: insights into the pathophysiology of *S. epidermidis* biofilms and the role of phenol-soluble modulins in formation of biofilms" in **J Infect Dis**, 191:289-298

Chapter 7

Clinical relevance and transcriptome studies of Pseudomonas aeruginosa biofilms

Susana P. Lopes, Ana M. Sousa, Diana Alves, Paula Jorge, Maria Olívia Pereira*
CEB - Centre of Biological Engineering, LIBRO – Laboratório de Investigação em Biofilmes Rosário Oliveira, University of Minho, Braga, Portugal
*** mopereira@deb.uminho.pt**

Contents of chapter

Introduction to the chapter

The ability of bacteria to survive in specific habitats and/or conditions, often unfavorable, requires the expression of thousands of environmentally regulated genes. The complexity of these regulatory networks increases the breadth of environments that bacterial species can occupy. Species that are able to survive in a broad range of environments generally possess a large coding capacity and devote a significant part of their genome to signal transduction and gene regulation.

Pseudomonas aeruginosa is ubiquitous in nature and is also an opportunistic human pathogen. It is also able to grow in a huge range of inhospitable environments, such those found in hospital settings, due to its persistence and great ability to form hardy biofilms. Therefore, infections caused by *P. aeruginosa* in hospitalized patients are common, being often severe and life-threatening, especially in patients with defects in host defenses (Costerton and Stewart, 1999; Lyczak and Cannon, 2002).

The main reason why *P. aeruginosa* is so harmful in the clinical setting is the fact that they are highly adaptable (Stover et al, 2000). This adaptive hallmark is a response to the environmental signals received that are relayed, often to regulators that modulate gene expression.

This chapter aims to present a brief overview about the clinical relevance of *P. aeruginosa* biofilms, highlighting the main infections caused by this opportunistic human pathogenic bacterium, and to review the transcriptome studies that have led to the advance of the understanding of the expression of the genes that control the diverse array of aspects underlying the pathogenesis of *P. aeruginosa*, namely the exacerbation of planktonic hallmarks, increased interspecies and inter-kingdom cell-cell signaling, biofilm formation and antimicrobial tolerance.

Pseudomonas aeruginosa

P. aeruginosa is a versatile organism that can survive in soil, marshes, and marine habitats, on plant and animal tissue, and on nonliving surfaces. It is also an increasingly prevalent opportunistic human pathogen.

The complete genome of this microorganism was firstly achieved in 2000, being among the largest genomes in the prokaryotic world, formed by 6.3 million base pairs (Stover et al, 2000). The enormous size of its genome enables *P. aeruginosa* to thrive in diverse environments, to infect a large variety of body sites and even to resist (intrinsically or by acquisition of new genes) to a large number of antimicrobial agents.

P. aeruginosa genome encodes an unusually high proportion of proteins involved in regulation, transport and virulence functions, providing a great deal of useful information, concerning not only the pathogenicity but also the potential of *P. aeruginosa* for resistance. Indeed, 0.3% of the total genes code for proteins are involved in antimicrobial resistance. It is well recognized that *P. aeruginosa* presents intrinsic resistance to a large number of antibiotics, given its low permeability of the outer-membrane, the over-expression of various efflux pumps, and the production of antibiotic target-inactivating enzymes (reviewed in Mesaros et al, 2007). In addition, *P.*

aeruginosa presents also a remarkable ability to develop or acquire new mechanisms of antimicrobial resistance. This may be related to the large size and the plasticity (or versatility) of its genome, which could constitute a reservoir for the species carrying other resistance genes. Frequently, both mechanisms (intrinsic and acquired) are present together, thereby conferring multi-drug resistant (MDR) phenotypes to *P. aeruginosa* (McGowan, 2006). The genome is also highly flexible, with 10% of genes organized in "pathogenicity islands", comprising variable genes coding for virulence factors, and with the ability to easily acquire large mobile genetic elements (integrons) encoding resistance genes (reviewed in Mesaros et al, 2007).

Most *P. aeruginosa* strains involved in infections are both invasive and toxigenic, due to the production of surface virulence factors (allowing bacterial adhesion colonization and tissue invasion) and secreted virulence factors (responsible for damage tissues or trigger the production of cytokines) (reviewed in Kipnis, 2006). In general, the specific syndromes accompanying an infection are a result of the combination of those virulent determinants expressed by each causal strain.

However, *P. aeruginosa* is able to modulate its virulence in different clinical contexts (depending on the infection site), revealing that this bacterium can easily adapt to the microenvironment encountered within the host by modulating the expression of many genes (Holban et al, 2013).

Impact of *P. aeruginosa* in clinical settings

In clinical scenery, this microorganism is persistent and easily forms biofilms, colonizing the human body and being consistently associated with severe and life-threatening infections (table 7.1). In this context, biofilms are extremely difficult to eradicate because of the limited susceptibility to antimicrobial agents and the high frequency of emergence of antibiotic resistance during therapy (Mah and O´Toole, 2001), thus resulting in severe adverse outcomes.

In the healthcare environment, *P. aeruginosa* is a leading cause of hospital-acquired (nosocomial) infections (American Thoracic Society, 2005), being estimated to be involved in 10-22.5% of such infections both in adults and in children. It is also responsible for high rates of illness and death mainly in immunocompromised individuals (Simonetti et al, 2013).

In recent years, hospital-acquired infections associated with indwelling devices have gained significant attention due to the increasing usage of implantable biomaterials coupled with the rise of resistant bacteria. The bacterial colonization of indwelling devices followed by biofilm formation remains a serious concern in modern healthcare as it is commonly associated to persistent infections called biomaterial-associated infections (BAI).

BAI infections can be originated from microorganisms found in the skin of patients and clinicians, in the water, in equipment that was not properly cleaned and sterilized or other sources in the environment (reviewed in Vertes et al., 2012). However, in device hospital-acquired infections *P. aeruginosa* stands out for its incidence and the ability to form strong biofilms on the surfaces of biomaterials.

Table 7.1 | Examples of some pathologies caused by *P. aeruginosa* biofilms (adapted from Mesaros et al, 2007)

Infection site	Specific associated pathologies	Occurrence or Incidence (%)
Blood	Sepsis (bacteraemia and septicaemia)	Frequent (51.7%)
Bone and joint infections	Osteochondritis of the foot Vertebral osteomyelitis Stenoarticular pyoarthrosis	Rare
Central nervous system	Meningitis Brain abscess	Rare
Ear	External otitis Chronic otitis media	Frequent
Eye	Keratitis Neonatal ophtalmia	Rare
Gastrointestinal tract	Necrotising enterocolitis Peri-rectal infections	Rare
Heart	Endocarditis	Rare (2.3%)
Nasal sinuses	Sinusitis	Frequent
Respiratory tract	Ventilator-associated pneumonia Cystic fibrosis Bronchiectasis Chronic obstructive pulmonary disease	Frequent (19%)
Skin	Burn and wound infections Dermatitis Folliculitis	Relatively frequent
Urinary tract	Acute and chronic urinary infections	Relatively frequent (17.2%)

Some examples of medical indwelling devices frequently colonized by this bacterium and the diseases or complications associated are presented in table 7.2.

Table 7.2 | Examples of diseases and other consequences of *P.aeruginosa* biofilms formed on indwelling medical devices
(adapted from Gottenbos et al., 2002 and Roosjen et al, 2006)

Indwelling medical device	Associated disease or consequences	Incidence (%)
Urinary catheter	Catheter-associated bacteriuria Blockage from thick films	10-20
Contact lenses	Keratitis Peripheral ulcer	0.6
Central venous catheter	Septicemia Endocartitis Blockage from thick films	4-12
Artificial hip prosthesis	Septicemia Prosthesis failure	2.6-4
Endotracheal tubes	Ventilator-associated pneumonia	9-18
Vascular graft	Infection	1-6
Wound dressings	Healing complications Chronic infection	1-5

Biofilm formation makes BAI extremely difficult to eradicate (reviewed in del Pozo and Patel, 2007) and for this reason, the fate of an infected implant device is often its surgical removal, at the expenses of considerable costs and patient suffering (reviewed in Gottenbos et al, 2002).

Ventilator-associated pneumonia (VAP) is one of the most frequent hospital-acquired infections that derived from endotracheal tubes contaminated by bacteria (e.g. *Staphylococcus aureus*, *P. aeruginosa* and *Enterobacteriaceae*). In this group of patients, *P. aeruginosa* is one of the most common and lethal pathogens colonizing the tubes, with attributable death rates ranging between 10 and 40% (Kollef, 2013). Acute *P. aeruginosa* infections, such as VAP, are invasive, cytotoxic, and frequently result in systemic infection, septic shock, and mortality. Therefore, the pathogenesis of *P. aeruginosa* in VAP infection is similar to that of acute infections, being multifactorial and dependent upon expression of many virulence determinants (e.g. secreted factors: elastase, phospholipase C, lecitinase, rhamnolipids; secondary metabolites: pyocyanin, cyanide), many of them regulated by the quorum-sensing (QS) system (Schuster and Greenberg, 2006). Furthermore, the emergence of *P. aeruginosa* strains, which exhibit co-resistance to almost all antibiotic classes, is increasing over the past decade (reviewed in Strateva and Yordanov, 2009). The presence of MDR strains in VAP patients diminishes the treatment options and increases the risk of inadequate empirical therapy, with *P. aeruginosa* causing early mortality (Peña et al, 2013). It has been suggested that *P. aeruginosa* or eventually other MDR species are “high-risk” pathogens and that patients who develop VAP due to these organisms are at high risk for poor clinical outcomes.

In addition to being a leading cause of nosocomial infections, such as VAP, this pathogen is also the main cause of chronic lungs infections including cystic fibrosis (CF) (reviewed in Coutinho et al, 2008). CF is an autosomal recessive lung disease caused by a defect in the cystic fibrosis conductance regulator (CFTR) gene. The lack of CFTR expression causes a defective chloride secretion creating an osmotic gradient that, consequently, provokes water hyper-reabsorption and abnormal thick and sticky sputum. The accumulated sputum is rich in nutrients being, thus, a good environment for microbial colonization (reviewed in Folkesson et al, 2012). The presence of *P. aeruginosa* in CF airways is highly associated with poor lung function, morbidity and mortality of patients. Despite the inflammatory response and the long-term and intensive antibiotic treatments, infections caused by *P. aeruginosa* persist in CF lungs. Once entering in CF airways, *P. aeruginosa* is virtually impossible to eradicate due to its remarkable genome plasticity that allows it to rapidly adapt to the greatly stressful CF environment (reviewed in Folkesson et al, 2012 and Hoiby et al, 2010). Its long persistence in CF airways is associated with sophisticated mechanisms of adaptation, including biofilm formation, resistance to antibiotics, hypermutability and customized pathogenicity in which virulence factors are expressed according the infection progression (reviewed in Sousa and Pereira, 2014).

Airway CF biofilms are genetic, proteomic and physiologic different of surface-attached biofilms formed, for instance, on indwelling devices (catheters, prostheses, pacemakers, stents) and medical and clinical equipment (Bjarnsholt et al, 2009). Instead of the direct surface colonization, clearly observed in device-associated infections, bacteria in the CF lungs preferentially form multicellular clusters or macrocolonies within the sputum and not on the surface epithelium in the bronchi and non-respiratory bronchioles as initially supposed (Bjarnsholt et al, 2009, Worlitzsch et al, 2002). Additionally, the environment in which CF-associated biofilms are formed is considered to be microaerophilic or anaerobic. The oxygen-limited and anaerobic growth conditions significantly increase antibiotic resistance of biofilm-forming bacteria (reviewed in Hassett et al, 2009).

P. aeruginosa causes other complication besides pulmonary and biomaterial related infections. Some common pseudomonal infections comprehend: soft tissue infections, related to risk factors such as burns, open wounds and postsurgery; bacteremia, usually associated with immunocompromised patients; diabetic foot, common in patients with diabetes and/or impaired microvascular circulation; otitis externa (swimmer's ear), associated with tissue injury or/and water blockage in ear canal; and otitis media folliculitis (hot tub rash), normally caused by improperly cleaned hot tubs (Gellatly and Hancock, 2013).

The above mentioned infections relate to the fact that *P. aeruginosa* is an opportunistic member of the skin microflora of both animals and humans (Hillier et al, 2008) and it possesses cell-associated virulence factors, such as pili, that facilitate colonization by binding to target host cell receptors. Pili's binding activates other cell wall adhesion proteins, resulting in permanent binding, which in turn stimulates the secretion of many other virulence factors by the bacterium. This "avalanche-like" increase in virulence factors is a unique offensive strategy of *P. aeruginosa* (Andonovaa and Uronova, 2013). Other characteristics, such as type IV pili, that also allows movement of the bacteria in order to find the most vulnerable site for penetration, and the activity of proteases (elastase B, elastase A, protease IV, alkaline protease), help *P. aeruginosa* in its successful skin infection pathogenesis (reviewed in Hoge et al, 2010).

Many wound related infections can acquire a chronic nature, especially when related with *P. aeruginosa* biofilm formation. Biofilms in wounds have been considered an obstacle of the natural healing process, since biofilms act as mechanical barriers, impeding re-epithelization, stimulating a chronic state of inflammation, and they provide protection from endogenous and exogenous antimicrobial agents (James et al, 2008; Bjarnsholt et al, 2008). Chronic wounds have a considerable load on healthcare costs. For example, chronically infected diabetic foot ulcers are considered the most significant wound care problem in the United States, leading many times to amputations, and the cost of caring for them is currently measured in the billions of dollars (Cunningham, 2006).

P. aeruginosa transcriptomic studies

The number of therapeutic options to treat infections caused by *P. aeruginosa* has significantly decreased in the last years. A key factor, in the search for novel approaches to treat or prevent these infections, is to know more about its biology. A thorough understanding of the transcriptomics of this bacterium is, therefore, crucial for finding the molecular basis of infections and, in particular, the mechanics of survival as biofilms (Waite et al, 2006).

P. aeruginosa in VAP

The presence of *P. aeruginosa* in the respiratory tract can be fatal, since this species presents an arsenal of virulence mechanisms that are well described as mandatory to poor prognosis in VAP. Three main virulence determinants have been described for *P. aeruginosa* against airway epithelia: the QS, the T3SS and lipopolysaccharide.

The production of virulence factors by *P. aeruginosa* is coordinated by QS, namely by the two main signaling systems, las and rhl (Le Berre et al, 2011). A recent study from Li and colleagues (2014) investigated the correlation between the expression of the las and rhl QS systems and the communal physiology of *P. aeruginosa* isolated from patients with hospital-acquired pneumonia. They observed that both QS genes, in particular the las signalling system, were strongly associated with bacterial motility, biofilm formation capacity, pyocyanin production and resistance to antibiotic treatment. Other study conducted by Khöler et al (2010) reported that VAP occurred more frequently in patients colonised by QS-proficient isolates, and that the production of rhamnolipids (controlled by the rhl system) by colonising isolates is potentially associated with the development of VAP in intubated patients.

The T3SS has four effector proteins (or exotoxins) - Exo S, Exo T, Exo U and Exo Y – and its production, together with elastase and pyocyanin in strains, have been linked to lung injuries (Roy-Burman et al, 2001). Between these Exo effectors, Exo U and S have been strongly related with increased virulence and with a risk of bacteremia during VAP (Le Berre et al, 2011). In addition, elastase was associated with acute lung injury and Exo U was associated (Le Berre et al, 2011).

Elastase exotoxin A is a potent toxin that contributes to the severity of *P. aeruginosa* lung infection, and its presence on the lungs results in the apoptotic and non-apoptotic cell death through some pathways (Lau et al, 2005).

P. aeruginosa in CF lungs

Transcriptomics has provided great inputs for drawing an evolutionary trajectory of *P. aeruginosa* within CF airways. During the development of infection, genotypes and phenotypes differ markedly from those that initially colonized CF airways. Driven by the challenging selective pressures imposed by the typical CF conditions, (e.g. interspecies competition, deficient oxygen, availability, biofilm growth, the immune system action, oxidative stress, and antibiotic treatment) *P. aeruginosa* gradually shifts from an acute virulent pathogen of early infection to a host-adapted pathogen of chronic infection (reviewed in Folkesson et al, 2012 and Ciofu et al, 2012).

Several transcriptomic studies have profiled isolates of *P. aeruginosa* in attempt to find the common route towards the chronic phenotype and the mechanisms underlying such a route. In particular, longitudinal studies have provided relevant information regarding the genetic changes undergo by *P. aeruginosa* and allowed comparing the expression of specific set of genes among patients in different periods of time. Gene expression changes in multidrug efflux pumps and regulators of QS and alginate biosynthesis have been identified, being the two latter hotspots of mutations (Huse et al, 2010; Smith et al, 2006).

QS is mediated by two types of molecules, *N*-acylhomoserine lactones (AHL) and 4-quinolones, allowing bacteria to perceive their density and regulate their gene expressions properly (reviewed in Hassett et al., 2010). Acute CF isolates produced AHL suggesting that QS circuit plays a role for *P. aeruginosa* pathogenesis at early stage of infection. Afterwards, QS seems be no longer needed and *lasR* mutants are frequently isolated in chronic stages. *lasR* gene encodes QS transcriptional regulator LasR and its downregulation may explain the reduced or absent production of AHL at this infection stage, the autolysis and the iridescent gloss of *P. aeruginosa* colonies, the growth advantage on amino acids and decreased virulence potential (Smith et al., 2006; D'Argenio et al, 2007).

Alginate overproduction confers the mucoid appearance to *P. aeruginosa* colonies, typically observed in chronic infection stages. Within the mucoid form, *P. aeruginosa* is more difficult to eradicate because it is highly resistance to antibiotics, as well as to the actions of host immune defenses (reviewed in Hassett et al, 2009). Alginate promotes *P. aeruginosa* encapsulation and biofilm formation protecting sessile bacteria from the action of reactive oxygen species, antibiotics and host immune defenses persisting in CF lungs (reviewed in Hoiby et al, 2010). The genetic mechanisms underlying *P. aeruginosa* transition to the mucoid form have been intensively studied and conversion is mainly caused by mutational inactivation of the *mucA* gene and rarely of *mucB* or *mucD* genes (reviewed in Deretic et al, 1995). *mucA* gene encodes a cytoplasmatic membrane bound protein that acts as anti-σ-factor, σ^{22}, limiting the expression of the *algD* operon required for alginate synthesis. σ^{22} can also activate the transcription of several other genes related to virulence factor expression and to stress response, including heat shock, and osmotic and oxidative stress (Wood and Ohman 2012) and repress the expression of type III secretion system (T3SS) genes (Wu et al, 2004). This suggests an impressive coordination of two high-cost energy systems in order to bacteria persist in CF airways.

The whole adaptation process to CF airways can be accelerated by the emergence of mutators phenotypes (or hypermutable phenotypes) which have high mutation rates up to 1000–fold than non-mutator phenotypes (Mena et al, 2008). In extreme selective conditions, such as those occurring in CF airways, this sophisticated mechanism improves the microevolution of *P. aeruginosa* accelerating its intraclonal diversification. The genes mainly affected are the antimutator genes *mutS*, *mutL*, and *uvrD* but it can be observed defects, as well in the genes *mutT*, *mutM*, and *mutY* (Ciofu et al, 2010). Mutators can increase the transcription of genes involved in the metabolism of fatty acids and amino acids crucial for obtaining energy in CF ecological niches where aerobic respiration is not possible (Hoboth et al, 2009).

P. aeruginosa in BAIs

A genotypic and phenotypic study of *P. aeruginosa* isolates from urinary tract and catheter-associated urinary tract infections demonstrated a heterogeneous in vitro production of several known virulence factors. However, factors such as the production of pyoverdin, protease and phospholipase A in addition to quorum sensing activity and biofilm formation were found in most of the isolates. On the other hand, other factors such as swarming phenotypes, the production of pyocyanin, phospholipase C and elastase were rarely observed. Strains isolated from catheter-associated infections showed significantly enhanced biofilm formation, decreased motility and slightly increased expression of virulence factor when compared to isolates from acute urinary tract infections (Tielen et al, 2010).

P. aeruginosa in chronic and skin related infections

A common mechanism of sensing external environmental signals is through two-component-regulatory systems (Stock et al, 2000). Goodman et al (2004) has showed that one of these regulatory systems, the RetS (regulator of exopolysaccharide and T3SS), are involved in sensing signals in the host and inversely controls the expression of genes associated with acute and chronic infection. The same authors further performed genome-wide transcriptional profiling to compare the *retS* mutant to the parent, where they showed that 397 in a total of 5678 *P. aeruginosa* genes analysed were regulated by RetS. Whilst virulence genes, with known roles in acute infection, such as the T3SS genes, were downregulated in the *retS* mutant, other genes (e.g. exopolysaccharide biosynthetic genes) were activated. It was supposed that the attenuated virulence, in the eukaryotic cells, showed by the *retS* mutant was consistent with the loss of T3SS, whereas the activated genes were responsible for biofilm formation on both surfaces and are potentially involved in the pathogenesis of chronic infections.

Recently, Turner and co-workers (2014) provided some insights about the genetic requirements for acute and chronic *P. aeruginosa* wound infections using both gene expression and fitness profiling. For that purpose, *P. aeruginosa* was grown *in vitro* and in two non-diabetic murine wound infection models (acute and chronic). Results showed no correlation between gene expression and mutant fitness for most genes, except for metabolic genes.

Recurrent *P. aeruginosa* ear infections in healthy divers led to a study that used cDNA microarray analysis to identify differentially expressed genes associated with virulence gene regulation by comparing the transcriptional profile of *P. aeruginosa* in a hyperbaric oxyhelium environment with a normal growth environment. In a hyperbaric oxyhelium environment, *P. aeruginosa* showed an upregulation of most virulence genes, such as stress-sense/response, protein folding, transcriptional regulation, pili and flagellum metabolism, virulence adaptation, and membrane protein metabolism (Chen et al, 2012).

P. aeruginosa adaptation to temperature is key to its pathogenesis upon infection of its mammalian host. A recent study examined the effects of growth temperature on the transcriptome of the *P. aeruginosa* PAO1. Microarray analysis of PAO1, grown at 22 °C and 37 °C, revealed a differential transcriptional regulation of 6.4 % of the genome. Major alterations were observed in metabolism, replication, and nutrient acquisition. For example, the expression of *P. aeruginosa* siderophores correlated to an increase in the extracellular concentration of Fe^{2+} at 37 °C and of Fe^{3+} at 22 °C. This suggests a change in the iron acquisition strategies when the bacterium switches from the external environment to the host (Barbier et al, 2014).

Acknowledgments

SL (SRFH/BPD/95616/2013), AMS (SFRH/BD/72551/2010), DA (SFRH/BD/78063/2011) and PJ (SFRH/BD/88192/2012) holds individual FCT fellowships.

References

American Thoracic Society; Infectious Diseases Society of America (2005) "Guidelines for the management of adults with hospital-acquired, ventilator-associated, and healthcare-associated pneumonia" in **American Journal of Respiratory and Critical Care Medicine**, 171:388-416

Andonova M, Urumova V (2013) "Immune surveillance mechanisms of the skin against the stealth infection strategy of *Pseudomonas aeruginosa* - Review" in **Comparative Immunology, Microbiology and Infectious Diseases**, 36:433–448

Bains M, Fernández L *et al.* (2012) "Phosphate Starvation Promotes Swarming Motility and Cytotoxicity of *Pseudomonas aeruginosa*" in **Applied and Environmental Microbiology**, 78: 6762–6768

Barbier M, Damron FH *et al.* (2014) "From the environment to the host: re-wiring of the transcriptome of *Pseudomonas aeruginosa* from 22°C to 37°C" in **PloS One**, 9:e89941

Bjarnsholt T, Kirketerp-Moller K *et al.* (2008) "Why chronic wounds will not heal: a novel hypothesis." In **Wound Repair and Regeneration**, 16:2–10

Bjarnsholt T, Jensen PO *et al.* (2009) "*Pseudomonas aeruginosa* biofilms in the respiratory tract of cystic fibrosis patients" **Pediatric pulmonology**, 44:547-558

Chen SH, Chen RY *et al.* (2012) "Microarray analysis and phenotypic response of *Pseudomonas aeruginosa* PAO1 under hyperbaric oxyhelium conditions" in **Canadian Journal of Microbiology**, 58:158–169

Ciofu O, Mandsberg LF *et al.* (2010) "Genetic adaptation of *Pseudomonas aeruginosa* during chronic lung infection of patients with cystic fibrosis: Strong and weak mutators with heterogeneous genetic backgrounds emerge in mucA and/or lasR mutants" in **Microbiology**, 156:1108–1119

Ciofu O, Mandsberg LF *et al.* (2012) "Phenotypes selected during chronic lung infection in cystic fibrosis patients: Implications for the treatment of *Pseudomonas aeruginosa* biofilm infections" in **FEMS Immunology and Medical Microbiology**, 65:215–225

Coutinho HD, Falcão-Silva VS, Goncalves GF (2008) "Pulmonary bacterial pathogens in cystic fibrosis patients and antibiotic therapy: a tool for the health workers" in **International Archives of Medicine**, 1:24

Costerton JW, Stewart PS (1999) "Bacterial biofilms: a common cause of persistent infection" in **Science**, 284:1318-1322

Cunningham AB (2006) "Biofilms: The Hypertextbook." in **Montana State University**, Bozeman

D'Argenio DA, Wu M *et al.* (2007) "Growth phenotypes of *Pseudomonas aeruginosa* lasR mutants adapted to the airways of cystic fibrosis patients" in **Molecular Microbiology**, 64:512–533

del Pozo JL, Patel R (2007) "The challenges of treating biofim-associated bacterial infections" in **Nature**, 82:204-209

Deretic V, Schurr MJ *et al.* (1995) "*Pseudomonas aeruginosa*, mucoidy and the chronic infection phenotype in cystic fibrosis" in **Trends in Microbiology**, 3:351–35

Folkesson A, Jelsbak L *et al.* (2012) "Adaptation of *Pseudomonas aeruginosa* to the cystic fibrosis airway: An evolutionary perspective" in **Nature Reviews Microbiology**, 10:841–851

Gellatly SL, Hancock REW (2013) "*Pseudomonas aeruginosa*: New insights into pathogenesis and host defenses" in **Pathogens and Disease**, 67:159–173

Goodman AL, Kulasekara B *et al.* (2004) "A signaling network reciprocally regulates genes associated with acute infection and chronic persistence in *Pseudomonas aeruginosa*", in **Developmental Cell**, 7:745-754

Gottenbos B, Busscher HJ *et al.* (2002) "Pathogenesis and prevention of biomaterial centered infections" in **Journal of materials science: Materials in medicine**,13:717-722

Hassett DJ, Korfhagen TR *et al.* (2010) "*Pseudomonas aeruginosa* biofilm infections in cystic fibrosis: insights into pathogenic processes and treatment strategies" **Expert Opinion Therapy Targets**, 14:117-130

Hassett DJ, Sutton MD *et al.* (2009) "*Pseudomonas aeruginosa* hypoxic or anaerobic biofilm infections within cystic fibrosis airways" in **Trends in Microbiology**, 17:130–138

Hillier A, Alcorn JR *et al.* (2008) "Pyoderma caused by *Pseudomonas aeruginosa* infection in dogs: 20 cases" in **Veterinary Dermatology**, 17(6):432–9

Hoboth C, Hoffmann R *et al.* (2009) "Dynamics of adaptive microevolution of hypermutable *Pseudomonas aeruginosa* during chronic pulmonary infection in patients with cystic fibrosis" in **The Journal of Infectious Diseases**, 200:118–130

Hoge R, Pelzer A *et al.* (2010) "Weapons of pathogen. Proteases and their role in virulence of *Pseudomonas aeruginosa*" in **Current research technology and education topics in applied microbiology and microbial biotechnology, FORMATEX**, 2:383–95

Hoiby N, Ciofu O *et al.* (2010) "*Pseudomonas aeruginosa* biofilms in cystic fibrosis" in **Future Microbiology**, 5:1663–1674

Holban AM, Chifiriuc MC *et al.* (2013) "Virulence markers in *Pseudomonas aeruginosa* isolates from hospital-acquired infections occurred in patients with underlying cardiovascular disease" in **Romanian Biotechnological Letters** 18:8843-8854

Huse HK, Kwon T *et al.* (2010) "Parallel evolution in *Pseudomonas aeruginosa* over 39,000 generations in vivo" in **mBio**, 1:e00199-10

Jabalameli F, Mirsalehian A *et al.* (2012) "Evaluation of biofilm production and characterization of genes encoding type III secretion system among *Pseudomonas aeruginosa* isolated from burn patients" in **Burns**, 38:1192-1197

James GA, Swogger E *et al.* (2008) "Biofilms in chronic wounds" in **Wound Repair and Regeneration**, 16:37–44

Kipnis E (2006) "Targeting mechanisms of *Pseudomonas aeruginosa* pathogenesis" in **Médecine et Maladies Infectieuses**, 36:78–91

Köhler T, Guanella R *et al.* (2010) "Quorum sensing-dependent virulence during *Pseudomonas aeruginosa* colonisation and pneumonia in mechanically ventilated patients" in **Thorax**, 65:703-710

Kollef MH (2013) "Ventilator-associated complications, including infection-related complications: the way forward" in **Critical Care Clinics,** 29:33-50

Lau GW, Hassett DJ *et al.* (2005) "Modulation of lung epithelial function by *Pseudomonas aeruginosa*" in **Trends in Microbiology,** 13:389-397

Le Berre R, Nguyen S *et al.* (2011) "Relative contribution of the main virulence factors in *Pseudomonas aeruginosa* pneumonia" in **Critical Carw Medicine,** 39:2113-2120

Li YH, Tian X (2012) "Quorum sensing and bacterial social interactions in biofilms" in **Sensors**, 12:2519–2538

Li Y, Qu H *et al.* (2014) "Correlation between group behavior and quorum sensing in *Pseudomonas aeruginosa* isolated from patients with hospital-acquired pneumonia", in **Journal of Thoracic Disease**, 6:810-817

Lyczak JB, Cannon CL (2002) "Lung infections associated with cystic fibrosis" in **Clinical Microbiology Reviews**, 15:194-222

Mah TF, O'Toole GA (2001) "Mechanisms of biofilm resistance to antimicrobial agents" in Trends in Microbiology, 9:34-39

McGowan JE Jr (2006) "Resistance in nonfermenting gram-negative bacteria: multidrug resistance to the maximum" in **American Journal of Infection Control** 34:S29-S37

Mena A, Smith EE *et al.* (2008) "Genetic adaptation of *Pseudomonas aeruginosa* to the airways of cystic fibrosis patients is catalyzed by hypermutation" in J **Journal of Bacteriology**, 190:7910–7917

Mesaros N, Nordmann P *et al.* (2007) "*Pseudomonas aeruginosa*: resistance and therapeutic options at the turn of the new millennium" in **European Society of Clinical Microbiology and Infectious Diseases**, 13:560-578

Peña C, Gómez-Zorrilla S *et al.* (2013) "Impact of multidrug resistance on *Pseudomonas aeruginosa* ventilator-associated pneumonia outcome: predictors of early and crude mortality" in **European Journal of Clinical Microbiology and Infectious Diseases**, 32:413-420

Rodrigue A, Quentin, Y *et al.* (2000) "Two-component systems in *Pseudomonas aeruginosa*: why so many?" in **Trends in Microbiology**, 8:498-504

Rodriguez-Rojas A, Oliver A *et al.* (2012) "Intrinsic and environmental mutagenesis drive diversification and persistence of *Pseudomonas aeruginosa* in chronic lung infections" in **The Journal of Infectious Diseases**, 205, 121–127

Roosjen A, Norde W *et al.* (2006) "The Use of Positively Charged or Low Surface Free Energy Coatings versus Polymer Brushes in Controlling Biofilm Formation" in **Progress in Colloid and Polymer Science**, 132:138-144

Roy-Burman A, Savel RH *et al.* (2001) "Type III protein secretion is associated with death in lower respiratory and systemic *Pseudomonas aeruginosa* infections" in **The Journal of Infectious Diseases,** 183:1767-1774

Schuster M, Greenberg EP (2006) "A network of networks: quorum-sensing gene regulation in *Pseudomonas aeruginosa*" in **International Journal of Medical Microbiology**, 296:73-81

Sharma G, Rao S *et al.* (2014) "*Pseudomonas aeruginosa* biofilm: Potential therapeutic targets" in **Biologicals**, 42:1–7

Simonetti A, Ottayano E *et al.* (2013) "Epidemiology of hospital-acquired infections in an adult intensive care unit: results of a prospective cohort study" in **Annals of Hygiene**, 25:281-289

Smith EE, Buckley DG *et al.* (2006) "Genetic adaptation by *Pseudomonas aeruginosa* to the airways of cystic fibrosis patients" in **Proceedings of the National Academy of Sciences of the United States of America**, 103:8487–8492

Sousa AM, Pereira MO (2014) "*Pseudomonas aeruginosa* Diversification during Infection Development in Cystic Fibrosis Lungs - a review" in **Pathogens**, 3:680-703

Stock AM, Robinson VL *et al.* (2000) "Two-component signal transduction" in **Annual Review of Biochemistry**, 69:183–215

Stover CK, Pham XQ *et al.* (2000) "Complete genome sequence of *Pseudomonas aeruginosa* PA01, an opportunistic pathogen" in **Nature,** 406:959– 964

Strateva T, Yordanov D (2009) "*Pseudomonas aeruginosa* - a phenomenon of bacterial resistance" in **Journal of Medical Microbiology**, 58:1133-1148

Tielen P, Narten M *et al.* (2010) "Genotypic and phenotypic characterization of *Pseudomonas aeruginosa* Tielen isolates from urinary tract infections" in **International Journal of Medical Microbiology**, 301:282-292

Turner K H, Everett J *et al.* (2014) "Requirements for *Pseudomonas aeruginosa* Acute Burn and Chronic Surgical Wound Infection" in **PLOS Genetics**, 10:1-12

Vertes A, Hitchins V *et al.* (2012) "Analytical challenges of microbial biofilms on medical devices" in **Analytical chemistry**, 84:3858-3866

Waite R D, Paccanaro A *et al.* (2006) "Clustering of *Pseudomonas aeruginosa* transcriptomes from planktonic cultures, developing and mature biofilms reveals distinct expression profiles" in **BMC Genomics**, 7:1-14

Wood LF, Ohman DE (2012) "Identification of genes in the σ^{22} regulon of *Pseudomonas aeruginosa* required for cell envelope homeostasis in either the planktonic or the sessile mode of growth" in **mBio**, 3:e00094-12

Worlitzsch D, Tarran R *et al.* (2002) "Effects of reduced mucus oxygen concentration in airway *Pseudomonas* infections of cystic fibrosis patients" **J Clin Invest**, 109:317-325

Wu W, Badrane H *et al.* (2004) "MucA-mediated coordination of type III secretion and alginate synthesis in *Pseudomonas aeruginosa*" in **Journal of Bacteriology**, 186:7575–7585

Chapter 8

Role of E. coli on catheter-associated urinary tract infections

Andreia S. Azevedo[1], Carina Almeida[1,2], Luís F. Melo, Nuno F. Azevedo[1]*

[1]Laboratory for Process Engineering, Environment, and Energy and Biotechnology Engineering (LEPABE), Department of Chemical Engineering, Faculty of Engineering, University of Porto, Rua Dr. Roberto Frias, 4200-465 Porto, Portugal

[2] CEB - Centre of Biological Engineering, LIBRO - Laboratory of Research in Biofilms Rosário Oliveira, University of Minho, Campus de Gualtar, 4710-057 Braga, Portugal

*** nazevedo@fe.up.pt**

Contents of chapter

Introduction to the chapter

Escherichia coli is a member of the *Enterobacteriaceae* family. This bacterium is a Gram-negative, facultative anaerobic, and non-sporulating with rod-shaped cells. As a commensal bacterium, *E. coli* is one of the most predominant microorganisms in the human fecal flora (reviewed in Probert and Gibson, 2002) but, this bacterium may also have a pathogenic role, causing numerous human infections, including urinary tract infections (UTIs), sepsis, meningitis, and diarrhea (reviewed in Probert and Gibson, 2002). The *E. coli* pathogenic nature is due to the presence of virulence factors that are responsible for toxins and siderophores production, adhesive and motility organelles, and resistance to antibiotics and host immune defense (Hacker and Kaper, 2000; Nataro and Kaper, 1998; reviewed in Oelschlaeger, 2002).

UTIs are the most common hospital-acquired (nosocomial) infections (Klevens, 2007) and can be defined when patients have a positive urine culture of $\geq 10^5$ CFU/ml ($\geq 10^4$ CFU/mL in children) in association with other symptoms (e.g. dysuria, urgency, fever) (Burckhardt and Zimmermann, 2011). The incidence of these infections increases with the use of indwelling devices, such as urinary catheters, which has been known as catheter-associated UTIs (CAUTIs) (reviewed in Curtis, 2008). The CAUTIs result from microbial colonization, with subsequent biofilm development and growth, on the catheter surface. In this form of life, bacteria are protected against the hydrodynamic forces caused by the urine flow, host defenses and antibiotics (reviewed in Warren, 2001).

Although CAUTIs are usually asymptomatic, the infection is inevitable for long-term urinary catheterization, with severe complications in most patients (reviewed in Jacobsen, 2008).

In the last decades, the study of the bacterial colonization process in medical devices (particularly in urinary catheters) has been receiving more attention. In the present chapter, we will provide information about the role of *E. coli* in CAUTIs development. In particular, we will discuss some transcriptomic and gene expression studies on *E. coli* biofilms in the scope of the CAUTIs.

Impact in clinical settings

CAUTIs represent the most common type of nosocomial infection (40% of all nosocomial infections in hospitals and nursing homes), account for over 1 million cases annually, in the United States, and constitute 80% of all nosocomial UTIs (reviewed in Stamm, 1991; reviewed in Warren, 1991). It has been estimated that about 25% of hospitalized patients will have a urinary catheter during their hospital stay and that the risk of developing an infection increases 5% with each day of catheterization (Bouza, 2001; Maki and Tambyah, 2001; reviewed in Warren, 2001). In addition to the higher risk of infection, CAUTIs are usually harder to treat due to the presence of crystalline biofilms (that further protect the microorganisms from treatments), which commonly contain multidrug resistance bacteria (reviewed in Jacobsen, 2008; Maki and Tambyah, 2001). Due to these features, the overall medical costs of nosocomial UTIs is staggering, with approximately $424 to $451 million spent annually, in the United States, to manage these infections (Tambyah, 2002). Also, catheter-associated bacteremia is estimated to cost approximately $2900 per episode (Saint, 2000) (including antimicrobial therapy, increased length of hospitalization, physician visits, and morbidity) and these costs are expected to rise due to advances in preventive medicine that extend life expectancy.

Previous studies reported that *E. coli* is responsible for 70%-95% of UTIs (reviewed in Kucheria, 2005) and is isolated from the urine of 20%-50% patients with urinary catheters (Macleod and Stickler, 2007; reviewed in Nicolle, 2005). The *E. coli* strains that are able to colonize and persist in the urinary tract are known as uropathogenic *E. coli* strains (UPEC).

Why *E. coli* is one of the most frequent microorganisms present in UTIs and CAUTIs remains unknown. Some authors have pointed out some explanations such as the presence of adhesins, toxins and flagellum-mediated motility (and other virulence factors) that are responsible for *E. coli* ability to invade and colonize, or the close proximity of the rectal area from the urinary tract that facilitates the bacteria dispersal (reviewed in Jacobsen, 2008). Recently, it was demonstrated that *E. coli* have a great ability to form a biofilm in conditions mimicking CAUTIs, whatever the preexisting microbiota, which helps explain the high prevalence of *E. coli* in CAUTIs (Azevedo, 2014). While these facts might partially explain the high prevalence of this species in CAUTIs, a more detailed analysis and identification of *E. coli* virulence factors would probably provide new insights for the development of novel strategies to prevent or minimize bacterial adhesion to catheters.

Catheter-associated urinary infections pathogenesis

The urinary catheters are inserted through the urethra into the bladder, to prevent/control the urinary retention and incontinence in hospitalized patients and in patients receiving home care (reviewed in Hall-Stoodley, 2004). The risk of CAUTIs usually depends on the catheterization time. For instance, patients undergoing short-term catheterization (up 7 days) are less susceptible to CAUTIs. However, in patients with long-term urinary catheters (several months or years) the infection is considered inevitable (reviewed in Warren, 1991).

In addition to *E. coli*, other bacteria from the endogenous intestinal flora (such as *Enterobacter*, *Klebsiella*, *Enterococci* or *Proteus*) are common colonizers of urinary catheters. Inadequately decontaminated equipment or improper practices of the healthcare workers, can also introduce environmental and common skin bacteria during the handling of urinary catheters. Thus, bacteria common on the hospital environments, such as *Pseudomonas*, *Serratia*, *Staphylococci* and *Acinetobacter*, can also cause healthcare-associated CAUTIs (Maki and Tambyah, 2001). Initially catheters are colonized by single pathogens (e.g. *E. coli*, *Enterococcus faecalis*), but later mixed biofilms may develop due to the adhesion of other species (e.g. *Proteus mirabilis*, *Providencia stuartii*, and *Klebsiella pneumonia*) that take advantage of the first colonizers to take over the catheter surface (reviewed in Jacobsen, 2008; Matsukawa, 2005).

Bacteria can infect the patients with urinary catheters by different ways: (i) bacteria can be introduced at the time of the catheter insertion, (ii) bacteria present in the urethra can migrate around the catheter extraluminally, or (iii) bacteria can ascend intraluminally through the lumen of the catheter as a result of the urine contamination in the drainage bag (reviewed in Stamm, 1991; Tambyah, 1999). In fact, the intermittent flow of warm nutritious urine on the surface of the urinary catheter offers attractive conditions for bacterial colonization (Ganderton, 1992).

The first step involved in the formation of catheter-associated biofilms is always the deposition of a conditioning film on the catheter surface. After the insertion of the urinary catheter in the patient, components of human urine (proteins, electrolytes and other organic molecules) can physico-chemically bind to the surface within minutes. These components are adsorbed rapidly onto the surface material and a conditioning film is formed (reviewed in Tenke, 2006). The conditioning film alters the surface characteristics of catheters, which facilitates the surface colonization and might also allow the adhesion of some microorganisms that do not have the ability to adhere on the catheter surface. This initial bacterial adhesion is reversible. However, an irreversible adhesion, with a strong anchoring of the microorganisms to the catheter, occurs due to the polysaccharides produced by bacteria (reviewed in Busscher, 2008; reviewed in Tenke, 2006). At this point, the bacterial phenotype changes and the development of a 3-dimensional biofilm structure takes place by the continuous division of bacteria and the production of extracellular polymeric substances (EPS) (reviewed in Tenke, 2006). Once the biofilm is established, the bacterial cells are protected from antibiotics agents and host immune defense mechanisms (reviewed in Tenke, 2006). In the case of UTIs, cells and aggregates of cells might detach and, when motile, can move against the flow allowing for the colonization of the bladder, kidneys and, eventually, the bloodstream.

A frequent problem associated to CAUTIs is the formation of crystalline deposits on the catheter surface, leading to trauma in the bladder and in the urethral epithelia. Also, the crystalline material (e.g. struvite, hydroxyapatite) on the luminal surface can lead to urine flow blockage in the catheter, aggravating the incontinence problem and facilitating the pathogen dispersion to the host tissues (reviewed in Stickler, 2008)

Currently, CAUTIs treatment or prevention relies on the prescription of antibiotics. The recurrent failure of antimicrobial agents (or other chemical agents) to prevent or treat the infection (reviewed in Tenke, 2006), has created the need for new effective control strategies. The prevention is predominantly based on enforcing guidelines for appropriate catheter placement (e.g. sterility during catheter handling and placement) or reduced catheterization time (reviewed in Boyce and Pittet, 2002; reviewed in Griffiths and Fernandez, 2007). Nonetheless, the incorporation of antimicrobial agents (e.g. nitrous oxide) on the urinary catheter surface (Regev-Shoshani,

2010) and the gradual release of antimicrobials from the catheters (reviewed in Hamill, 2007), have been proposed as alternative ways to reduce the uropathogens adhesion and biofilm formation. The use of nonpathogenic "probiotic" microorganisms, bacteriophages or inhibitors of quorum sensing (Bala, 2011; Carson, 2010; Trautner, 2007) have also been regarded as efficient alternatives to treat or prevent CAUTIs.

E. coli in catheter-associated urinary infections transcriptome studies

To better understand the virulence of *E. coli* strains associated to CAUTIs, several transcriptome studies have been developed. By studying the crucial steps involved in biofilm formation, key factors implicated in the adhesion, biofilm development and invasion process by UPEC strains, have been identified. In table 8.1, we present the most important virulence factors implicated in UPEC persistence and invasion process during CAUTIs development.

To start the infection, the UPEC have to bind to the urinary catheter surface, therefore, these microorganisms must be capable of expressing and producing specific adhesion factors, namely adhesins. These adhesins are a well-known group of virulence factors expressed by UPEC during the course of the infection that are involved in their initial attachment to the uroepithelial cells and to the urinary catheter surface. The type 1 fimbriae is the most frequently expressed adhesin in UPEC (reviewed in Jacobsen, 2008).

Table 8.1 | Examples of virulence factors that have been implicated in the persistence and invasion of UPEC strains, during CAUTIs development.

Adhesins	
Type 1 fimbrae	Encoded by the *fim* gene. Recognize epithelial cells, immune cells, extracellular matrix proteins and erythrocytes.
P pili	Encoded by the papABCDEFGHIJK operon; it is the second most common virulence factor expressed by UPEC, but their role in CAUTIs is unclear.
Dr adhesines family	Includes the fimbrial adhesin Dr and nonfimbrial adhesins, and are involved in the invasion process of epithelial cells; but their expression during CAUTIs is unknown.
Motility	
Flagella	Might provide the ability to disseminate from the catheter surface to the urinary tract; but there is no evidence of its importance for virulence of UPEC strains.
Persistence, biofilm formation and acquisition of nutrients	
Toxins	The cytotoxic necrotizing factor 1 (CNF1) has been shown to have a role in the UPEC invasion into uroepithelial cells; The hemolysins induce the cell lysis.
Capsular polysaccharide	Avoid the host immune response; but there are no studies that confirm its role in CAUTIs.
Lipo-polysaccharide	A constituent of the gram-negative cell wall which might be important in antibiotic resistance.
Iron acquisition systems	Encoded by genes (e.g. *entC, iroB, sitA, iucD*) responsible for the production of molecules capable to sequestrate the iron present in the urine, namely siderophores (e.g. enterobactin, aerobactin).
Other virulence factors	Encoded by genes responsible for the production of antibiotics resistance systems (e.g. *marAB*, *yhjX* and *emrA* genes), and the nitrogen limitation adaptation (e.g. *glnA, glnp, gadA, gadB* genes)

A previous study demonstrated that the type 1 fimbriae is associated with the *E. coli* persistence in the patients undergoing long-term urinary catheterization (Mobley, 1987). In a study by Snyder and co-workers (2004), the analysis of infected mice with *E. coli* CFT073, has shown that type 1 fimbriae was highly upregulated (Snyder, 2004). However, later Ross and Klemm (2006) analyzed the expression profile of *E. coli* strain 83972 from three patients with UTI and showed that the fimbrial adhesins were not upregulated (Roos and Klemm, 2006). Regarding these two studies, it is important to keep in mind that strain CFT073 and strain 83972 have some differences which could explain the results obtained in their expression profiles. The *E. coli* CFT073 has been described as a pathogenic organism; while *E. coli* 83972 has been labelled as non-virulent/asymptomatic and in fact can prevent the infection by pathogenic bacteria. Therefore, this inability to produce an aggressive phenotype might be partially explained by the absence of a functional fimbriae. Other surface adhesins including S pili (encoded by *sfa* operon), F1C pili (encoded by an operon of six genes *focAICDFGH*), F9 fimbriae (encoded by *f9* operon), IrgA adhesin (encoded by *iha*), Dr adhesins (encoded by *draA, draC, draD, draE* genes) or P fimbriae (encoded by *pap* operon) are frequently present in UPEC strains (reviewed in Mulvey, 2002). While these adhesins are likely to be involved in the adhesion process to the surface or tissue, it is also thought that they have a limited role during CAUTIs caused by UPEC. For instance, P fimbriae are the second most common virulence factors associated with UPEC but, while some studies have shown their involvement in the UTI development, others have failed to show this requirement (reviewed in Svanborg and Godaly, 1997; Tseng, 2001). The expression of adhesion factors by UPEC in the urinary tract seems well established. However, there is limited information on how these factors interfere with the UPEC adhesion to catheters surface.

Another important factor for UPEC attachment on the surface catheter and further dispersal of the biofilm cells is motility. Despite the lack of evidences, the flagellum-mediated motility might allow the movement of microorganisms on the catheter surface and from the catheter to other sites, such as the bladder, ureter, kidney, and uroepithelium (reviewed in Jacobsen, 2008; reviewed in Mobley, 1994). In fact, flagellum-mediated motility is conserved among UPEC strains, with 70% of clinical isolates obtained from women with UTI demonstrating motility when grown in semisolid agar (Wright, 2005). Nevertheless, *in vivo* and *in vitro* data have demonstrated a downregulation of genes encoding flagellar proteins (e.g. *flgL*, *flgK*, *motAB*, *ycgR*, *yhiH*); therefore, there is no evidences supporting the importance of motility during UTIs (Snyder, 2004).

Once a mature biofilm develops on the surface of catheters, or on the uroepithelium, *E. coli* develops strategies to ensure its survival in a hostile environment. This is accomplished by the expression of genes responsible for the capsular polysaccharide and lipopolysaccharide (LPS) synthesis, iron acquisition systems, antibiotics resistance mechanisms, nitrogen limitation and oxygen levels adaptation (Snyder, 2004), and toxins production (e.g. hemolysins and cytotoxic necrotizing factor 1 [CNF1]) (reviewed in Jacobsen, 2008). In more detail, it has been shown that genes responsible for the capsule transportation (e.g. *kpsE, kpsD, kpsC, kpsS, nanA*), LPS (e.g. *lpxA, lpxB, wzzE*) synthesis, antibiotic resistance (e.g. *marAB, yhjK, emrA*), and the catabolism of alternative carbon sources (e.g. *uxuA, uxuB, uxaA* and *uxaB*) are upregulated (Snyder, 2004). In the case of the capsular polysaccharide, it has been suggested that these genes may be involved in UPEC virulence by avoiding the immune response, conferring resistance to serum and phagocytosis (Horwitz and Silverstein, 1980; reviewed in Jacobsen, 2008; reviewed in Johnson, 2003). In terms of biofilms, the capsule has been shown to have an indirect role in biofilms by shielding bacterial surface adhesins (Schembri, 2004). Specifically, the group II capsule seems to display anti-adhesion properties, by weakening cell-surface contacts (in initial bacterial adhesion) and reducing cell–cell interactions when the biofilm matures (Valle, 2006). On the other hand, colanic acid, which forms a protective capsule around the bacterial cell under specific growth and environmental conditions, is induced by near-lethal levels of a subset of β-lactam antibiotics that may exacerbate the formation and persistence of a biofilm (Sailer, 2003). Although colanic acid has also been reported to impair initial attachment, its synthesis is consistently upregulated within biofilms, and this compound plays a role in the development of the mature biofilm architecture (reviewed in Beloin, 2008).

Concerning the upregulation of genes responsible for the LPS biosynthesis, it has been shown that mutations affecting LPS synthesis influence the *E. coli* ability to adhere to abiotic surfaces (Genevaux, 1999), suggesting a role on the adhesion process. Similarly to what happen with the capsule, some authors also suggest that the LPS masks the function of short membrane adhesins. A reduction in LPS expression might unmask the *E. coli* adhesion allowing its initial adhesion or biofilm formation (reviewed in Beloin, 2008).

Regarding the expression of antibiotic resistance genes during UTIs, it is unclear why, in the absence of the antibiotic, there is an induction of the resistance genes. In fact, biofilms resistance is commonly associated with the matrix presence, that imposes a diffusion barrier and the low metabolic state of the biofilm cells that hinders the uptake of the drug (reviewed in Costerton, 1999). However, it has been documented that SOS responses, typically induced during biofilm development, might induce the bacteria resistance mechanisms (Dorr, 2009; reviewed in Poole, 2012). This means that the upregulation of these genes might be the result of a coordinated adaptive response to an environmental stress that aims at creating a highly adapted hypermutated phenotype. For the catabolism of alternative carbon sources, *in vivo* experiments have shown that in the urinary tract metabolism shifted so that hexuronates, hexanates, glucitol and fructose could be exploited to support growth (Snyder, 2004).

Concerning the toxins, including hemolysin and CNF1, they provide UPEC with the means to inflict extensive tissue damage, facilitating bacterial dissemination as well as releasing host nutrients and disabling immune effector cells (reviewed in Wiles, 2008). Nonetheless, there are inconsistent results. For instance, regarding CNF1, Hertting et al. (2008) showed that the there was no difference between *E. coli* strains from children with sporadic infections and the group of children with UTI (Hertting, 2008). Regarding microcins, which are secreted antibiotic peptides, the upregulation that was found to occur *in vivo* (Snyder, 2004), may be related to the microorganism ability to compete with surrounding bacteria.

The access to oxygen might be another concern for bacteria in biofilms or within the human body. *E. coli*, a facultative anaerobe, can adapt to growth in various redox environments. The *in vivo* studies showed that either genes encoding molecules involved in anaerobic growth (e.g. *fdnGHI* operon) or genes indicating the aerobic growth (e.g. *cyoAB*, *cyoE*, *cytochrome o*) were upregulated (Snyder, 2004), which is consistent with the presence of different microenvironments into the biofilm. Within biofims, it has been reported that anaerobic growth does not support biofilm formation (Colon-Gonzalez, 2004), but later other authors have shown that this behavior is highly strain dependent (reviewed in Bjergbæk, 2014).

In fact, it seems that the access to oxygen is very important for *E. coli* biofilm formation, since previous transcriptomic studies on *E. coli* biofilms have identified two upregulated gene clusters known to be induced under limiting oxygen conditions (Schembri, 2003). On the other hand, the nitrogen and iron are essential nutrients for the survival of UPEC in UTIs and CAUTIs. In fact, iron and nitrogen acquisition systems are considered important virulence factors during the CAUTIs and UTIs progression. The iron concentration is low in urine, but it is crucial for the microorganisms growth and survival during CAUTIs (reviewed in Jacobsen, 2008), being consumed by microorganisms with a high ability to produce or utilize siderophores. In fact, several studies have demonstrated that genes encoding the siderophores molecules were upregulated in strains isolated from the urinary tract and from infected mice (Demir and Kaleli, 2004; Snyder, 2004). These systems have also been shown to be important in biofilm formation (Hancock, 2010).

Additionally, nitrogen-regulated genes are highly expressed to maximize the amount of nitrogen assimilated, since *E. coli* lacks the urease enzyme required to convert the urea (present at high levels in urine) to ammonia and CO_2 (Hagan, 2010; Snyder, 2004). Glutamine synthetase is an important enzyme involved in glutamine synthesis and in ammonia metabolism; therefore the high expression of *glA* (encodes the glutamine synthetase), *glnP* and *glnQ* (encode glutamine importers) genes observed *in vivo* studies might be crucial for the *E. coli* strains when the nitrogen available during the infection is limited (Hagan, 2010; Snyder, 2004).

In short, the studies cited above have generated a list of potential virulence factors that might have an important role in *E. coli* strains colonization and infection. In general, most of the studies have analyzed the gene expression *in vivo* using *E. coli* strains from patients with UTIs; or *in vitro* using human urine to mimic the conditions found in

UTIs. However, further research on *E. coli* gene expression using the physiological conditions found in catheterized patients or using clinical samples derived from patients undergoing a CAUTI, is needed. In this way, a recent study has developed a dynamic catheterized bladder model for analyzing *in vitro* catheter colonization factors (Reisner, 2014). This will allow the combination of the generated data with *in vivo* analysis of clinical samples from catheterized patients. The authors have applied this model to further investigate the contribution of type 1 fiambriae to catheter colonization, and concluded that type 1 fiambriae are important for UPEC catheter colonization *in vitro*, and were also highly expressed in urine samples from catheterized patients. This study shows the benefits of using dynamic models for studying the virulence factors involved in catheter colonization (Reisner, 2014). The continuous identification and analysis of genes encoding for virulence factors and its coordinated expression, might provide the development of new tailored strategies to prevent and control CAUTIs.

Acknowledgments

ASA and CA hold an individual FCT fellowship (SFRH/BD/82663/ 2011 and SFRH/BPD/74480/2010 respectively.)NFA is an Investigator FCT.

References

Azevedo AS, Almeida C *et al.* (2014) "Interaction between atypical microorganisms and *E. coli* in catheter-associated urinary tract biofilms" in **Biofouling**, 30:893-902

Bala A, Kumar R *et al.* (2011) "Inhibition of quorum sensing in *Pseudomonas aeruginosa* by azithromycin and its effectiveness in urinary tract infections" in **J Med Microbiol**, 60:300-306

Beloin C, Roux A *et al.* (2008) "*Escherichia coli* biofilms" in **Curr Top Microbiol Immunol**, 322:249-289

Bjergbæk LA, Haagensen JAJ *et al.* (2014) "Effect of oxygen and growth medium on in vitro biofilm formation by *Escherichia coli*." in **Biofilms**, 3:1-10

Bouza E, San JR *et al.* (2001) "A European perspective on nosocomial urinary tract infections II. Report on incidence, clinical characteristics and outcome (ESGNI-004 study). European Study Group on Nosocomial Infection" in **Clin Microbiol Infect**, 7:532-542

Boyce JM and Pittet D (2002) "Guideline for Hand Hygiene in Health-Care Settings. Recommendations of the Healthcare Infection Control Practices Advisory Committee and the HICPAC/SHEA/APIC/IDSA Hand Hygiene Task Force. Society for Healthcare Epidemiology of America/Association for Professionals in Infection Control/Infectious Diseases Society of America" in **MMWR Recomm Rep**, 51:1-45, quiz CE 1-4

Burckhardt I and Zimmermann S (2011) "*Streptococcus pneumoniae* in urinary tracts of children with chronic kidney disease" in **Emerg Infect Dis**, 17:120-122

Busscher HJ, Norde W *et al.* (2008) "Specific molecular recognition and nonspecific contributions to bacterial interaction forces" in **Appl Environ Microbiol**, 74:2559-2564

Carson L, Gorman SP *et al.* (2010) "The use of lytic bacteriophages in the prevention and eradication of biofilms of *Proteus mirabilis* and *Escherichia coli*" in **FEMS Immunol Med Microbiol**, 59:447-455

Colon-Gonzalez M, Mendez-Ortiz MM *et al.* (2004) "Anaerobic growth does not support biofilm formation in *Escherichia coli* K-12" in **Res Microbiol**, 155:514-521

Costerton JW, Stewart PS *et al.* (1999) "Bacterial biofilms: a common cause of persistent infections" in **Science**, 284:1318-1322

Curtis LT (2008) "Prevention of hospital-acquired infections: review of non-pharmacological interventions" in **J Hosp Infect**, 69:204-219

Demir M and Kaleli I (2004) "Production by *Escherichia coli* isolates of siderophore and other virulence factors and their pathogenic role in a cutaneous infection model" in **Clin Microbiol Infect**, 10:1011-1014

Dorr T, Lewis K *et al.* (2009) "SOS response induces persistence to fluoroquinolones in *Escherichia coli*" in **PLoS Genet**, 5:e1000760-

Ganderton L, Chawla J *et al.* (1992) "Scanning electron microscopy of bacterial biofilms on indwelling bladder catheters" in **Eur J Clin Microbiol Infect Dis**, 11:789-796

Genevaux P, Bauda P *et al.* (1999) "Identification of Tn10 insertions in the *rfaG*, *rfaP*, and *galU* genes involved in lipopolysaccharide core biosynthesis that affect *Escherichia coli* adhesion" in **Arch Microbiol**, 172:1-8

Griffiths R and Fernandez R (2007) "Strategies for the removal of short-term indwelling urethral catheters in adults" in **Cochrane Database Syst Rev**,CD004011-

Hacker J and Kaper JB (2000) "Pathogenicity islands and the evolution of microbes" in **Annu Rev Microbiol**, 54:641-679

Hagan EC, Lloyd AL *et al.* (2010) "*Escherichia coli* global gene expression in urine from women with urinary tract infection" in **PLoS Pathog**, 6:e1001187-

Hall-Stoodley L, Costerton JW *et al.* (2004) "Bacterial biofilms: from the natural environment to infectious diseases" in **Nat Rev Microbiol**, 2:95-108

Hamill TM, Gilmore BF *et al.* (2007) "Strategies for the development of the urinary catheter" in **Expert Rev Med Devices**, 4:215-225

Hancock V, Dahl M *et al.* (2010) "Abolition of biofilm formation in urinary tract *Escherichia coli* and *Klebsiella isolates* by metal interference through competition for fur" in **Appl Environ Microbiol**, 76:3836-3841

Hertting O, Chromek M *et al.* (2008) "Cytotoxic necrotizing factor 1 (CNF1) induces an inflammatory response in the urinary tract *in vitro* but not *in vivo*" in **Toxicon**, 51:1544-1547

Horwitz MA and Silverstein SC (1980) "Influence of the *Escherichia coli* capsule on complement fixation and on phagocytosis and killing by human phagocytes" in **J Clin Invest**, 65:82-94

Jacobsen SM, Stickler DJ *et al.* (2008) "Complicated catheter-associated urinary tract infections due to *Escherichia coli* and *Proteus mirabilis*" in **Clin Microbiol Rev**, 21:26-59

Johnson JR (2003) "Microbial virulence determinants and the pathogenesis of urinary tract infection" in **Infect Dis Clin North Am**, 17:261-78, viii

Klevens RM, Edwards JR *et al.* (2007) "Estimating health care-associated infections and deaths in U.S. hospitals, 2002" in **Public Health Rep**, 122:160-166

Kucheria R, Dasgupta P *et al.* (2005) "Urinary tract infections: new insights into a common problem" in **Postgrad Med J**, 81:83-86

Macleod SM and Stickler DJ (2007) "Species interactions in mixed-community crystalline biofilms on urinary catheters" in **J Med Microbiol**, 56:1549-1557

Maki DG and Tambyah PA (2001) "Engineering out the risk for infection with urinary catheters" in **Emerg Infect Dis**, 7:342-347

Matsukawa M, Kunishima Y *et al.* (2005) "Bacterial colonization on intraluminal surface of urethral catheter" in **Urology**, 65:440-444

Mobley HL, Chippendale GR *et al.* (1987) "Expression of type 1 fimbriae may be required for persistence of Escherichia coli in the catheterized urinary tract" in **J Clin Microbiol**, 25:2253-2257

Mobley HL, Island MD *et al.* (1994) "Virulence determinants of uropathogenic *Escherichia coli* and *Proteus mirabilis*" in **Kidney Int Suppl**, 47:S129-S136

Mulvey MA (2002) "Adhesion and entry of uropathogenic *Escherichia coli*" in **Cell Microbiol**, 4:257-271

Nataro JP and Kaper JB (1998) "Diarrheagenic *Escherichia coli*" in **Clin Microbiol Rev**, 11:142-201

Nicolle LE (2005) "Catheter-related urinary tract infection" in **Drugs Aging**, 22:627-639

Oelschlaeger TA, Dobrindt U *et al.* (2002) "Virulence factors of uropathogens" in **Curr Opin Urol**, 12:33-38

Poole K (2012) "Bacterial stress responses as determinants of antimicrobial resistance" in **J Antimicrob Chemother**, 67:2069-2089

Probert HM and Gibson GR (2002) "Bacterial biofilms in the human gastrointestinal tract" in **Curr Issues Intest Microbiol**, 3:23-27

Regev-Shoshani G, Ko M *et al.* (2010) "Slow release of nitric oxide from charged catheters and its effect on biofilm formation by *Escherichia coli*" in **Antimicrob Agents Chemother**, 54:273-279

Reisner A, Maierl M *et al.* (2014) "Type 1 fimbriae contribute to catheter-associated urinary tract infections caused by *Escherichia coli*" in **J Bacteriol**, 196:931-939

Roos V and Klemm P (2006) "Global gene expression profiling of the asymptomatic bacteriuria *Escherichia coli* strain 83972 in the human urinary tract" in **Infect Immun**, 74:3565-3575

Sailer FC, Meberg BM *et al.* (2003) "beta-Lactam induction of colanic acid gene expression in *Escherichia coli*" in **FEMS Microbiol Lett**, 226:245-249

Saint S (2000) "Clinical and economic consequences of nosocomial catheter-related bacteriuria" in **Am J Infect Control**, 28:68-75

Schembri MA, Dalsgaard D *et al.* (2004) "Capsule shields the function of short bacterial adhesins" in **J Bacteriol**, 186:1249-1257

Schembri MA, Kjaergaard K *et al.* (2003) "Global gene expression in *Escherichia coli* biofilms" in **Mol Microbiol**, 48:253-267

Snyder JA, Haugen BJ *et al.* (2004) "Transcriptome of uropathogenic *Escherichia coli* during urinary tract infection" in **Infect Immun**, 72:6373-6381

Stamm WE (1991) "Catheter-associated urinary tract infections: epidemiology, pathogenesis, and prevention" in **Am J Med**, 91:65S-71S

Stickler DJ (2008) "Bacterial biofilms in patients with indwelling urinary catheters" in **Nat Clin Pract Urol**, 5:598-608

Svanborg C and Godaly G (1997) "Bacterial virulence in urinary tract infection" in **Infect Dis Clin North Am**, 11:513-529

Tambyah PA, Halvorson KT *et al.* (1999) "A prospective study of pathogenesis of catheter-associated urinary tract infections" in **Mayo Clin Proc**, 74:131-136

Tambyah PA, Knasinski V *et al.* (2002) "The direct costs of nosocomial catheter-associated urinary tract infection in the era of managed care" in **Infect Control Hosp Epidemiol**, 23:27-31

Tenke P, Kovacs B *et al.* (2006) "The role of biofilm infection in urology" in **World J Urol**, 24:13-20

Trautner BW, Hull RA *et al.* (2007) "Coating urinary catheters with an avirulent strain of *Escherichia coli* as a means to establish asymptomatic colonization" in **Infect Control Hosp Epidemiol**, 28:92-94

Tseng CC, Huang JJ *et al.* (2001) "Decreased predominance of papG class II allele in *Escherichia coli* strains isolated from adults with acute pyelonephritis and urinary tract abnormalities" in **J Urol**, 166:1643-1646

Valle J, Da RS *et al.* (2006) "Broad-spectrum biofilm inhibition by a secreted bacterial polysaccharide" in **Proc Natl Acad Sci U S A**, 103:12558-12563

Warren JW (1991) "The catheter and urinary tract infection" in **Med Clin North Am**, 75:481-493

Warren JW (2001) "Catheter-associated urinary tract infections" in **Int J Antimicrob Agents**, 17:299-303

Wiles TJ, Kulesus RR *et al.* (2008) "Origins and virulence mechanisms of uropathogenic *Escherichia coli*" in **Exp Mol Pathol**, 85:11-19

Wright KJ, Seed PC *et al.* (2005) "Uropathogenic *Escherichia coli* flagella aid in efficient urinary tract colonization" in **Infect Immun**, 73:7657-7668

Chapter 9

Listeria monocytogenes: a major foodborne pathogen

Diana Rodrigues
CEB - Centre of Biological Engineering, LIBRO - Laboratory of Research in Biofilms Rosário Oliveira, University of Minho, Campus de Gualtar, 4710-057 Braga, Portugal
dianaalexandra@deb.uminho.pt

Contents of chapter

Introduction to the chapter

Listeria monocytogenes continues to be one of the major microorganisms responsible for severe food contamination worldwide leading to serious, and potentially fatal, food born diseases in human and animal. These bacteria can be present in many different types of food products – meat, dairy, seafood, vegetables -, and are especially found on raw or ready-to-eat products (reviewed in Valk, 2005; Mena, 2004). Moreover, post-process contamination of a food product with *L. monocytogenes* is an ongoing problem, because the pathogen can survive and grow at refrigeration temperatures and in environments with reduced water activity. Its capacity to adhere to several food contact materials such as stainless steel, polypropylene, and glass has also been documented (e.g.. Ronner, 1993) and, like many others, these bacteria have the ability to form biofilms and live as communities that confer protection face to stressful environmental conditions, as well as resistance to disinfectants and antimicrobial agents (e.g. Aarnisalo, 2000). Therefore, biofilms are considered one of the most relevant virulence factors of *Listeria* spp. and it is of utmost importance to study all the mechanisms and processes involved in their formation under as many different conditions as possible.

It is, then, essential to fully understand the molecular mechanisms on which stress adaptation in *L. monocytogenes* is based, in order to better restrain its propagation in food and thus decrease the occurrence of food-borne listeriosis. In this regard, transcriptomic studies have been performed on planktonic cells and biofilms under different environmental conditions, and their gene expression analyzed (e.g. Garmyn, 2012). In this chapter, it is presented an overview of the impact of *L. monocytogenes* on public health and its presence on food industry. Moreover, some significant transcriptomic studies and how they contributed to unravel virulence aspects of this pathogen are also presented.

Impact of listeriosis

L. monocytogenes bacterium is a facultative intracellular, gram-positive, and responsible for serious diseases that include meningitis, abortion, and septicaemia (reviewed in Vazquez-Boland, 2001). This pathogen passes through the different kinds of host's barriers (fetoplacental, gastrointestinal, blood-brain) and gets inside epithelial and phagocytic cells, where it replicates. The individuals that are more vulnerable to listeriosis are expectant women, foetuses, neonates, immunocompromised, and elderly people. This disease has a low infection rate but a high mortality rate (approximately 20%–30%), which makes *L. monocytogenes* a very serious worry for food industry and public health (Doyle, 2001).

The latest report of the European Food Safety Authority showed that the number of confirmed human listeriosis cases increased to 1,642, and that a total of 198 deaths, due to listeriosis, were reported by 18 European Member States in 2012, which was the highest number of fatal cases reported since 2006 (EFSA, 2014).

L. monocytogenes in food industry

The colonization of food-contact surfaces can lead to several problems in food industry, including economical losses and public health issues (e.g. Di Bonaventura, 2008). The lack of efficiency of cleaning and sanitation procedures allows microorganisms' adhesion and, oftentimes, the development of biofilms on such surfaces, which constitutes a potential source of food contamination. This biofilms can grow in pipes, drains, joints, water and sewage lines, heat exchangers, packaging surfaces, etc., developing in all the inert surfaces as plastic, stainless steel, glass, wood, rubber, formica, iron, marble, granite, among others (Teixeira, 2005; Dogan, 2003; Austin,1995). The presence of biofilms in food industry is particularly problematic in dairy industry, processing of meat, poultry and fish, brewing industry, and fresh processing industry (reviewed in Kumar, 1998). Here, just dairy, meat and fish industries will be highlighted, and only regarding the impact of *L. monocytogenes.*

In dairy industry, foodborne pathogens from milking systems (Teixeira, 2005), fecal origin, washwater from dairies or storage systems can reach the processing lines and contaminate the entire line (Jayarao, 2001), from the pipes and milk storage tanks to the milk itself and its derivates. Moreover, the contamination of dairy products already pasteurized takes place in fillers or due to biofilms present in the joints (Dogan, 2003; Austin, 1995). Although *L. monocytogenes* is not the bacterium most frequently found in dairy industry, it still represents a big challenge for those responsible for food safety because of its relatively high rate of mortality (Riemann, 2006). Strains of this pathogen have been repeatedly isolated from bulk milk storage tanks and present good biofilm formation ability, which results in the contamination of dairy products (Borucki, 2004). This fact is due to its psychrotrophic nature and its capacity to persist on surfaces for a long time. Food outbreaks caused by milk ingestion or dairy products seem to be, according to the specialists, caused by biofilms since they occur sporadically, not on a continuous basis. This happened, for example, with outbreaks caused by *L. monocytogenes* present in cheese (reviewed in Zottolla, 2001).

In meat industry, the danger of contamination begins immediately with the slaughter of animals. In this stage, the presence of faecal matter from the slaughtered animals can contaminate the meat that will be processed and packaged with high levels of bacteria, including *L. monocytogenes.* Various studies have demonstrated the presence of pathogenic microorganisms in meats. For example, samples of minced beef from meat industries were collected over the US and it was found that 11.7% was contaminated with *L. monocytogenes* (Schlosser, 2001). In terms of food safety in meat industry, this bacterium is the most problematic microorganism due to its capacity to survive, and even grow, at refrigeration temperatures or in meat products ready and packaged under vacuum or modified atmosphere (Jessen, 2003). In fact, it was verified that *L. monocytogenes* can be found in meat processing areas with a very high incidence (e.g. Wendlandt, 1994), and its survival – which may be years – makes possible that meat processing lines can be recurrently contaminated.

In fish industry, there are reports about the occurrence of persistent bacteria, indicating that microorganism in biofilms are important contaminants in these processing environments (reviewed in Shi, 2009). For example, there are data on the prevalence of *L. monocytogenes* in a tropical fish (*Pseudoplatystoma* sp.), smoked and minimally processed, that reveal that 5% of the units are contaminated with this bacterium, and has been demonstrated that the isolated strains belonged to the same subtype to those isolated in same factory four years before (Alves, 2005). Moreover, in a study about the microbial ecology of different fish processing industries it as reported that *Listeria* sp. was detected in all the facilities surveyed (Bagge-Ravn, 2003).

Transcriptomic studies about *L. monocytogenes* virulence

One of the main challenges, regarding microbial pathogenesis, is to unravel the molecular processes that allow the change from saprophytism to virulence. Although ubiquitous in nature, *L. monocytogenes* becomes highly problematic when infects a host, within which it becomes mostly intracellular (reviewed in Cossart, 2011). This alternative lifestyle is due to the ability of this bacterium to defy macrophage killing, and to its superb capacity to infect a diversity of non-phagocytic cells, such as epithelial cells of the intestine (reviewed in Hamon, 2006). In order to study the molecular mechanisms behind these *Listeria*'s virulence features, a mix of molecular biology, cell biology and classical genetic approaches have been used.

After the first complete genome sequence of *L. monocytogenes* in 2001 (Glaser, 2001), several other genomic studies have been performed with this bacterium (Nelson, 2004; Dussurget, 2002) which, by their turn, allowed to proceed with in-depth researches regarding gene expression. One of the first works in this area was focused on the *prfA* gene (Milohanic, 2003), which was confirmed to codify a protein essential to control very important virulence genes. Moreover, a main contribution of this work was the discovery of associations between *prfA* induction and the expression of *sigB*, the gene that encodes the sigma B regulon (Milohanic, 2003) which is one of five sigma factors in *Listeria.* During the last ten years, this regulon, together with regulons for other transcription regulatory factors, have been analysed using whole-genome transcriptional profiling. Accordingly, microarray analysis allowed concluding that the *sigB* expression is related with the control of over hundred genes (Hain, 2008; Raengpradub 2008). It was also found that some of these genes are related with virulence and response to stress, and are up-regulated at the stationary phase.

RNA-sequencing and genomic tilling arrays are also two potent methods that have been used for genome-wide transcriptome analysis (reviewed in Sorek, 2009). Some years ago, Oliver and coworkers (Oliver, 2009) used RNA-sequencing to study the *sigB* expression in the 10403S *L. monocytogenes* strain, which is one of the most widely used strains to explore the virulence of this pathogen. Regarding stationary phase, this study showed that 83% of annotated genes are transcribed in that phase, and 96 genes are over-expressed in the wild type comparing with a *sigB* mutant. Another study has used genomic-tilling array analysis to examine the transcriptomes of *Listeria* grown *in vitro* (under different growth phases, low oxygen concentration, and different temperatures), *in vivo* (in the intestine of germ-free animals), and *ex vivo* (in blood) (Toledo-Arana, 2009). This work was performed with the EGD-e strain (another strain widely used to study *Listeria* virulence) and its respective *prfA*, *sigB* and *hfq* mutants, which transcriptomes were analyzed and compared. The *hfq* gene corresponds to a RNA-binding protein, whose function is to stabilize RNA-RNA hybrids, and is also related with *Listeria* virulence (Christiansen, 2004). That work (Toledo-Arana, 2009) lead to the valuable conclusion that the expression of *prfA* gene controls the transcription of virulence genes in the blood, while *sigB* expression is related with the activation of virulence genes in the intestine.

The transcriptomic works performed regarding *L. monocytogenes* virulence have also been showing significant strain dissimilarities, which helps to elucidate the particular characteristics of a specific strain, and/or associate structural aspects with phenotypes. Nevertheless, this also makes it harder to establish common principles. Future transcriptomic studies, probably using a combination of different methodologies, are expected to reveal further information about *L. monocytogenes'* virulence and resistance.

Acknowledgments

DR holds an individual FCT fellowship (SFRH/BPD/72632/2010).

References

Aarnisalo K, Salo S *et al.* (2000) "Bactericidal efficiencies of commercial disinfectants against *Listeria monocytogenes* on surfaces" in **Journal of Food Safety**, 20:237-250

Alves VF, De Martinis EC *et al.* (2005) "Antilisterial activity of a *Carnobacterium piscicola* isolated from Brazilian smoked fish (surubim [*Pseudoplatystoma* sp.]) and its activity against a persistent strain of *Listeria monocytogenes* isolated from surubim" in **Journal of Food Protection**, 68(10):2068-2077

Austin JW, Bergeron G (1995) "Development of bacterial biofilms in dairy processing plants" in **Journal of Dairy Research**, 62:509–519

Bagge-Ravn D, Yin Ng *et al.* (2003) "The microbial ecology of processing equipment in different fish industries—analysis of the microflora during processing and following cleaning and disinfection" in **International Journal of Food Microbiology**, 87:239-250

Borucki MK, Reynolds J *et al.* (2004) "Dairy farm reservoir of *Listeria monocytogenes* sporadic and epidemic strains" in **Journal of Food Protection**, 67(11):2496-2499

Christiansen JK, Larsen MH *et al.* (2004) "The RNA-binding protein Hfq of *Listeria monocytogenes*: role in stress tolerance and virulence" in **Journal of Bacteriology**, 186:3355-3362

Cossart P (2011) "Illuminating the landscape of host-pathogen interactions with the bacterium *Listeria monocytogenes*" in **Proceedings of the National Academy of Sciences of the United States of America**, 108:19484–19491

Di Bonaventura G, Piccolomini R *et al.* (2008) "Influence of temperature on biofilm formation by *Listeria monocytogenes* on various food-contact surfaces: relationship with motility and cell surface hydrophobicity" in **Journal of Applied Microbiology**, 104:1552–1561

Dogan B, Boor KJ (2003) "Genetic diversity and spoilage potentials among *Pseudomonas* spp. isolated from milk products and dairy processing plants" in **Applied and Environmental Microbiology**, 69:130-138

Doyle MP, Beuchat LR, Montville TJ, editors (2001) "Food Microbiology: Fundamentals and Frontiers" 2nd ed. Washington, DC: American Society for Microbiology

Dussurget O, Cabanes D *et al.* (2002) "*Listeria monocytogenes* bile salt hydrolase is a PrfA-regulated virulence factor involved in the intestinal and hepatic phases of listeriosis" in **Molecular Microbiology**, 45(4):1095-106

EFSA (2014) "The European Union Summary Report on Trends and Sources of Zoonoses, Zoonotic Agents and Food-borne Outbreaks" in **EFSA Journal**, 12(2):3547

Garmyn D, Augagneur Y *et al.* (2012) "*Listeria monocytogenes* differential transcriptome analysis reveals temperature-dependent Agr regulation and suggests overlaps with other regulons" in **PLoS ONE**, 7(9):e43154

Glaser P, Frangeul L *et al.* (2001) "Comparative genomics of *Listeria* species" in **Science**, 294(5543):849-852

Hain T, Hossain H *et al.* (2008) "Temporal transcriptomic analysis of the *Listeria monocytogenes* EGD-e sigma B regulon" in **BMC Microbiology**, 8:20

Hamon M, Bierne H *et al.* (2006) "*Listeria monocytogenes*: a multifaceted model" in **Nature Reviews Microbiology**, 4(6):423-434

Jayarao BM, Henning DR (2001) "Prevalence of foodborne pathogens in bulk milk" in **Journal of Dairy Science**, 84:2157–2162

Jessen B, Lammert L (2003) "Biofilm and disinfection in meat processing plants" in **International Biodeterioration & Biodegradation**, 51:265-269

Kumar CG, Anand SK (1998) "Significance of microbial biofilms in food industry: a review" in **International Journal of Food Microbiology**, 42:9–27

Mena C, Almeida G *et al.* (2004) "Incidence of *Listeria monocytogenes* in different food products commercialized in Portugal" in **Food Microbiology**, 21:213–216

Milohanic E, Glaser P *et al.* (2003) "Transcriptome analysis of *Listeria monocytogenes* identifies three groups of genes differently regulated by PrfA" in **Molecular Microbiology**, 47(6):1613-1625

Nelson KE, Fouts DE *et al.* (2004) "Whole genome comparisons of serotype 4b and 1/2a strains of the food-borne pathogen *Listeria monocytogenes* reveal new insights into the core genome components of this species" in **Nucleic Acids Research**, 32(8):2386-2395

Oliver HF, Orsi RH *et al.* (2009) "Deep RNA sequencing of *L. monocytogenes* reveals overlapping and extensive stationary phase and sigma B-dependent transcriptomes, including multiple highly transcribed noncoding RNAs" in **BMC Genomics**, 10:641

Raengpradub S, Wiedmann M *et al.* (2008) "Comparative analysis of the sigma B-dependent stress responses in *Listeria monocytogenes* and *Listeria innocua* strains exposed to selected stress conditions" in **Applied and Environmental Microbiology**, 74:158-171

Riemann HP, Cliver DO (2006) "Foodborne Infections and Intoxications" 3rd ed. Amsterdam, the Netherlands: Elsevier Academic Press

Ronner AB, Wong ACL (1993) "Biofilm development and sanitizer inactivation of *Listeria monocytogenes* and *Salmonella typhimurium* on stainless steel and bruna-N rubber" in **Jounal of Food Protection**, 56:750-758

Schlosser E (2001) "Fast Food Nation" New York: Houghton Mifflin Company, p197

Shi X, Zhu X. (2009) "Biofilm formation and food safety in food industries" in **Trends in Food Science & Technology**, 20:407-413

Sorek R, Cossart P (2010) "Prokaryotic transcriptomics: a new view on regulation, physiology and pathogenicity" in **Nature Reviews Genetics**, 11: 9-16

Teixeira P, Lopes Z *et al.* (2005) "Physico-chemical surface characterization of a bacterial population isolated from a milking machine" in **Food Microbiology**, 22:247-251

Toledo-Arana A, Dussurget O *et al.* (2009) "The *Listeria* transcriptional landscape from saprophytism to virulence" in **Nature**, 459:950-956

Valk Hd, Jacquet C *et al.* (2005) "Surveillance of *Listeria* infections in Europe" in **Eurosurveillance**, 10:251- 255

Vazquez-Boland JA, Kuhn M *et al.* (2001) "*Listeria* pathogenesis and molecular virulence determinants" in **Clinical Microbiology Reviews**, 14:584-640

Wendlandt A, Bergann T (1994) "*Listeria monocytogenes*: occurrence in a factory for slaughtering, carving and meat processing" in **Fleischwirtsch**, 74:1329–1331

Zottolla EA (2001) "Reflections on *Salmonella* and other "wee beasties" in foods" in **Food Technology**, 55(9):60-67

Chapter 10

Streptococcus mutans biofilms: an insider's look

Eduardo M. Costa, Sara Silva, Manuela E. Pintado*
CBQF – Centro de Biotecnologia e Química Fina – Laboratório Associado, Escola Superior de Biotecnologia, Universidade Católica Portuguesa/Porto, Rua Arquiteto Lobão Vital, Apartado 2511, 4202-401, Porto, Portugal
*** mpintado@porto.ucp.pt**

Contents of chapter

Introduction to the chapter

The oral cavity is a complex system composed by soft and hard tissues were the capability to adhere is required for bacterial survival. At the present and with *ca.* 700 species of microorganisms already identified in the human oral cavity, among which several with pathogenic potential, one has been singled out as the central microorganism – *Streptococcus mutans* (Dong, 2012).

The first report of *S. mutans* comes from 1924 when Clarke isolated a bacterial species from carious lesions that on a Gram stain was more oval than round, thus appearing to be a mutant form of a coccus (Clarke, 1924). However, 35 years passed without a single reference to *S. mutans* until it was rediscovery in the 1960s when it was identified in rodent models and routinely associated with dental decay (Carlssson, 1968; Loesche, 1986, Banas, 2004).

Nowadays, *S. mutans* is widely recognized as being the primary etiological agent of human dental caries and an almost "obligate" biofilm forming bacteria (Wen, 2002). It is also recognized for its ability to survive and adapt to environmental threats, a key characteristic considering that ever evolving environment of the oral cavity (Bitoun, 2012). Furthermore, and despite not being labeled as a primary oral colonizer, *S. mutans* has a recognized role in the initiation of dental caries and is a key contributor to the development of the dental plaque (Costa, 2013). With dental caries being a biofilm-dependent chronic disease with high socioeconomic impact, it stands to reason that knowledge regarding biofilm formation mechanisms, of its primary etiological agent, be of the upmost importance.

Establishing a foothold

The adhesion of *S. mutans*, within dental plaque, can be mediated via sucrose-independent and dependent means. Sucrose-independent adhesion is thought to be profoundly influenced by a 185 kDa surface protein antigen known by various names, such as SpaP or antigen I/II. Encoded by the *pac* gene, it's reported to be involved in cellular hydrophobicity (Nakano, 2006). Mutants lacking this protein do not bind as well as the wild type to surfaces and show lower capability to induce caries in rats (Banas, 2004;).

On the other hand, sucrose dependent adhesion is characterized by the action of glucosyltransferases (GTFs) in the synthesis of glucans and is considered to be the major mechanism behind sucrose dependent adhesion (Shemesh, 2007). From a molecular standpoint, GTFs expression (*gtfB*, *gtfC* and *gtfD*) is regulated by the *vicRKX* operon, an operon that has been deemed as essential for *S. mutans* survival as it influences the expression of several virulence associated genes affecting synthesis of polysaccharides (such as GTFs) and polysaccharide-binding sites (Adil, 2014; Shemesh, 2007). From a biological standpoint, GTFs possess a sucrose activity that results in the splitting of sucrose, the only natural substract for the GTFs, into glucose and fructose. The glucose moiety is then added to a growing polymer of glucan. Collectively, GtfB, GtfC and GtfD synthesize both water soluble and water insoluble glucans, the first being predominantly a linear polymer linked by α-1,6-glycosidic linkages that resembles dextran and the second possessing a higher degree of branching and, predominantly, α-1,3-linkages (Shemesh, 2006). Both types of polymers are thought to contribute to sucrose-dependent colonization, but the water insoluble type may be of primary importance for smooth surface caries (Nakano, 2006). While they work as a collective GTFs, there are some intrinsical differences between them. GtfB and GtfC are located on the cell surface of *S. mutans* and synthesize water insoluble glucans; on the other hand GtfD is an extracellular enzyme, as it is released into culture supernatants and synthesizes water soluble glucans (Nakano, 2006; Shemesh, 2006).

Presumably, sucrose dependent adhesion can involve binding by glucan-coated bacteria, or attachment of *S. mutans* to glucan previously present in the dental plaque (Banas, 2004; Shemesh, 2006). One possibility for the presence of this glucan is the synthesis by extracellular GTFs that bound to the salivary pellicle. Another hypothesis is that it was synthesized by *S. mutans*, or other oral streptococci, that had previously adhered via sucrose-dependent means. The basis for glucan binding is still subject to speculation. While studies have confirmed the glucan-binding properties of GTFs, most of these proteins are exported and lack cell wall anchoring motifs. The glucan-binding domain (GBD) of the GTFs consists of a carboxyl terminal region of amino acid repeats that is shared by at least two non-enzymatic glucan bindingin proteins, GbpA and GbpD. It is possible that another non-enzymatic protein, GbpC, acts as a cell surface glucan receptor, though the aggregation phenotype attributed GbpC is only observed under certain stressful conditions. Another possibility is the wall-associated protein WapA. Although not previously described as a glucan-binding protein, the inactivation of the *wapA* gene resulted in a reduction in aggregation and adhesion (Banas, 2004). One less discussed route of adhesion is the usage of fructosyltransferase (Ftf) (expressed by the *ftf* gene) by *S. mutans*. This enzyme synthesizes fructan polymers from sucrose, which may act as binding sites for bacterial accumulation (Klein, 2009; Shemesh, 2006).

Growing and multiplying

Once colonization is achieved, the first step for *S. mutans* is building the colony, which means synthesis of the extracellular polysaccharide (EPS) matrix that will house the biofilm. In the presence of sucrose, *S. mutans* will produce EPS continuously resulting in a highly structure 3D matrix. However this is only possible due to the abundance of genes directly involved in EPS matrix synthesis (*gtfB*, *gtfC*, *gtfD*, *ftf*), remodelling (*dexA*, *fruA*) and regulation (*vicRKX*, *gcrG*, *luxS*, *rpoA*, *ccpA* and *manL*) (Klein, 2012). Once again, GTFs are fundamental in the assembly of the EPS matrix, as they produce the glucans, which are essential for the assembly and structural integrity of the matrix. During this phase, several regulatory systems are involved – VicRK (codified by the *vicRKX operon*) and LuxS (expressed by the *luxS* gene) in the regulation of GtfB and GtfC (expressed by the *gtfB* and *gtfC* genes) synthesis of glucan and the regulatory proteins CcpA and ManL (expressed by *ccpA* and *manL* genes), which are also involved in the up-regulation of EPS synthesis in the early stage of biofilm formation (Duque, 2011). Another two enzymes that are also noteworthy during this phase are dextranase (*dexA*) and Ftf (*ftf*). The first digests soluble glucans providing more substract for acid production and insoluble glucan synthesis and the fructans formed by the later provide storage of extracellular nutrients and help keep the biofilm hydrated (Klein, 2010; Mattos-Graner, 2004).

In mature biofilms the scenery is somewhat different. The first proof of alteration of genetic expression is the decline in GTFs production. Once the matrix is assembled there is a reduced need for glucan synthesis and, as such, these genes are either inhibited or repressed (Klein, 2012). Other alteration is the expression of the cell density signals such as de cell-density-dependent Com system, which will be fundamental in the formation of the biofilm and in the acid tolerance of the biofilm, and the LuxS enzyme (*luxS*), which affects biofilm accumulation and biofilm structure (Wen, 2004; Yoshida, 2005). Furthermore, the expression of global regulators such as Ccpa, RelA and RopA (expressed by the *ccpA*, *relA* and *ropA* gens), all of which play significant roles in biofilm formation, will increase in this phase (Merritt, 2003; Wen, 2006). However, all of this enzymes, proteins and regulatory signals are not as fundamental as the BrpA protein.

The biofilm regulatory protein A (BrpA) (expressed by the *brpA* gene) is a surface-associated protein that has been proven to be fundamental in biofilm formation. Mutant strains lacking this protein present severe defects in biofilm formation, autolysis and an overall attenuated virulence (Wen, 2002; Wen, 2006; Yoshida, 2002). Furthermore, strains lacking *brpA* fail to accumulate and form mature biofilms, despite being capable to adhere to surfaces. Another proof of *brpA* importance is its key role in stress tolerance as *brpA*-mutants show little growth in acidified media and an increased sensitivity to acid killing and hydrogen peroxide induced oxidative stress (Wen, 2006). Last, but not least, when comparing the genetic expression of a *brpA*-mutant with a wild type, the expression of over 100 proteins were altered in the absence of *brpA* and more than 41 gens were up/down-regulated by at least 1.5 fold (Wen, 2006). Among the down-regulated genes are those responsible for coding the IIBC domain of the fructose-specific phosphotransferase system, one of the primary enzymes responsible for fructose transport, and for galactokinase GalK (expressed by the *galK* gene), which is a part of the galactose metabolism. Amid the up-regulated genes are those responsible for encoding homologues of the manganese-type superoxide dismutase, a hydrogen peroxide-forming NADH oxidase and peroxidase resistance protein, which plays a key role in oxidative stress tolerance (Wen, 2002; Wen, 2006).

Feeling the consensus

Quorum Sensing (QS) generally describes the mechanisms used in bacterial intracellular communication, therefore it stands to reason that biofilms, as organized communities, are also coordinated by QS systems. The major QS system in *S. mutans* was discovered using as base the information of closely related *Streptococcus* species, whose QS system was relatively well characterized (Li, 2002; reviewed in Suntharalingam, 2005). While, in the beginning, it was believed that the QS system's purpose was to induce competence, thus facilitating the acquisition of exogenous DNA, it is now known that this system regulates several phenotypes besides competence induction such as biofilm formation, bacteriocin production and acid tolerance response (ATR) (Li, 2001; Li, 2002). The streptococci QS pathway, mediated by the competence stimulating peptide (CSP), encompasses at least two different genetic loci, *comCDE* and *comAB*, that encode the peptide signaling system (*comCDE*) and the specific secretion apparatus (*comAB*) (Li, 2001; Li, 2002; reviewed in Suntharalingam, 2005). The specific secretion apparatus is composed by an ATP-binding cassete transporter (ComA) and accessory protein (ComB), which compose a membrane system that is responsible for the release of CSP into the extracellular media. When the concentration of this compound reaches a threshold, CSP will interact with its specific receptor, a membrane-bound Histidine Kinase (HK) sensor protein (ComD) triggering a cascade of early, late and delayed gene induction and gene repression transcriptional profiles that are temporally distinct (reviewed in Suntharalingam, 2005).

The early gene transcriptional profile (early competence mechanism) is a cell to cell signaling apparatus that implies a peptide-signaling system. This mechanism is encoded in the *comCDE* operon and is comprised of a CSP precursor (ComC – expressed by the *comC* gene), the HK sensor protein (ComD – expressed by the *comD*) and a cognate response regulator – ComE (expressed by *comE* gene). The *comE* gene is responsible for the upregulation of the transcription of *comAB*, *comCDE* and *com*X genes. The ComX is an alternate sigma factor, whose production is stimulated during this stage, which will act as regulator and inducer, together with an RNA polymerase, of the late and delayed gene induction and repression (Suntharalingam, 2005).

One of the most fascinating findings, regarding *S. mutans* competence and its association with biofilms, was observed by Li (2001) which found that the genetic transformability of biofilm derived cells was 10 to 600 times higher in cells derived from biofilms than their planktonic counterparts (Li, 2001). This strengthened the association between competence and biofilms. In fact, when using *comC*, *comD*, *comE* and *comX* mutants, all the resulting biofilms exhibited an abnormal architecture (that compromised their integrity) with the *comD*, *comE* and *comX* mutants biofilms also exhibiting lower biomass levels than a wildtype control (Li, 2001; Li, 2002; reviewed in Suntharalingam, 2005). On the other hand, the addition of exogenous, synthetic CSP, will not only induce competence and increase transformation efficiency, but also significantly increase the amount of biofilm formed, thus further strengthening the link between biofilms, competence and the QS system (reviewed in Suntharalingam, 2005). Another evidence that supports this is the fact that biofilm embedded cells are much more effective in incorporating exogenous DNA than their planktonic counterparts (Li, 2001). This is a likely result of the upregulation of the *comCDE* operon in the CSP-QS system as the accumulation of CSP is relatively easier in biofilms than in liquid culture media and suggests that biofilm formation appears to favor optimal CSP-QS functioning and, therefore facilitates the horizontal exchange of genes (Li, 2001; reviewed in Suntharalingam, 2005).

In an attempt to further elucidate the relationship between the CSP-QS system and biofilm formation, the genetic expression of different *S. mutant* strains' biofilm (wild type and mutants deficient in the *comC*, *comD* or *comE* genes) was analyzed. Expression analysis showed that the ComE/ComD cascade exhorted a positive regulatory effect upon the expression of glucosyltransferases (*gtfB*, *gtfC*, *gtfD*), frucrosyltransferase (*ftf*) and a glucan binding protein (*gbpB*), with the glucosyltransferases using glucan to produce polymers that promote plaque formation (Senadheera, 2008).

Stress management

A key feature, necessary for *S. mutans* survival, is its capability to survive to external aggressions – either from environmental changes or antibiotic presence. Multiple pathways have been identified as being used by *S. mutans* to modulate its capacity to deal with stresses and their importance in maintaining biofilms has been described as being paramount (Bitoun, 2012).

When considering adapting and surviving in adverse environmental conditions, Bitoun (2012) showed that certain two-component signal transducing systems, including CiaHR (*ciaRH*), VicRK (*vicRKX* operon) and LiaSR (*liaSR*), play a fundamental role in *S. mutans* sensitivity to low pH and reactive oxygen. Additionally, *brpA* is involved in acid and oxidative stress tolerance responses and biofilm development by *S. mutans.* Mutant strains lacking this protein show an increased sensitivity to low pH and hydrogen peroxide (Bitoun, 2011). Another protein that seems to be fundamental is RopA. RopA is expressed by the gene *ropA* and is an apparent trigger factor (a ribosome-associated peptidyl-prolyl *cis-trans* isomerase, which is a molecular chaperone that is highly conserved) homologue. At low pH *ropA* absence seems to cause a decrease in survival rate of 45-fold compared to the wild type. Furthermore, the loss of RopA alters the expression of at least 33 genes, 22 of which were up-regulated and 11 down-regulated. Among the upregulated genes can be found *dnaK, groEL,* and *gorES,* molecular chaperones that are inducible by heat shock and acid stress and are believed to play a central role in the ability of oral bacteria to cope with frequent stresses encountered in the oral cavity (Bitoun, 2012). Considering that trigger factor in *Escherichia coli* is associated with GroEl/ES in promoting the affinity of these chaperones and binding capacity to various stress-damaged proteins, deficiency of RopA will likely diminish the capacity of the organism to deal with stresses imposed by low pH, oxygen and other detrimental conditions (Wen, 2005).

A more specialized, but nevertheless fundamental gene, is termed *rex* (for redox suppressor) and it responds directly to changes in cytoplasmatic NADH/NAD^+ ratio. The transcription of repressed *rex* has been shown to be activated in response to oxygen limitations (in *Bacillus subtilis*) which leads to production of cytochrome *bd* and NADH-linked lactate dehydrogenase, ensuring efficient oxygen utilization and recycling the excess of NADH. In *S. mutans*, the *rex* gene was found to be part of the *brpA*-regulon with significant decrease in a *brpA*-deficient mutant during the early exponential phase, suggesting that *rex* expression is influenced by *brpA* and that *rex* may be involved in the regulation of stress tolerance response and/or biofilm formation (Bitoun, 2011).

When considering resistance to antimicrobials, most resistance mechanisms seem to be centred in the cell envelope where two-component signal transduction systems (TCSTSs) are among the primary sensory-regulatory mechanisms that mediate bacterial adaptation processes. In gram positive bacteria this system may be activated via exposure to lipid II cycle inhibitors such as the antibiotics vancomycin and bacitracin, the bacteriocin nisin, and cationic antimicrobial peptides; hence, its cogname, lipid II interacting antibiotics LiaRS (Dong, 2012). In *S. mutans*, in particular, 13 more TCSTSs have been identified with 4 (ComDE, CiaRH, VicRK and LiaSR) being already known and shown to play a prominent role in regulating stress response. In fact, *liaSR* deficient mutants showed to be two to fourfold more sensitive to cell envelope biosynthesis target antimicrobials than the wild type (Bitoun, 2012; Dong, 2012; Suntharalingam, 2009).

Conclusions

Overall, *Streptococcus mutans* genetic machinery seems to be completely geared towards the biofilm lifestyle. Described by some as the perfect colonizer, *S. mutans* is capable of expressing enzymes and proteins perfectly tailored for each step of biofilm development and growth (as can be seen in table 10.1). Furthermore, its highly efficient *quorum sensing* system in conjunction with its innate capability to adapt to environmental changes and antibiotic presence makes *S. mutans* the ultimate colonizer of the oral cavity. As such, the affirmation of Dong (2012), which considered "*S. mutans* the root of all evil in the oral cavity", is not without merit as *S. mutans* seems to be geared towards colonizing and infecting the oral cavity and its presence is the catalyzer that will allow the subsequent colonization by other microorganisms.

Table 10.1 | Summary of genes involved in the stages of *S. mutans* biofilms.

Gene associated	Function
pac	Sucrose-independent adhesion – involved in cellular hydrophobicity
gtfB	Sucrose-dependent adhesion – synthesis of insoluble glucans EPS matrix synthesis
gtfC	Sucrose-dependent adhesion – synthesis of insoluble glucans EPS matrix synthesis
gtfD	Sucrose-dependent adhesion – synthesis of soluble glucans EPS matrix synthesis
gbpA	Sucrose-dependent adhesion – glucan-binding protein
gbpD	Sucrose-dependent adhesion – glucan-binding protein
gbpC	Sucrose-dependent adhesion – cell surface glucan receptor

wpaA	Sucrose-dependent adhesion – aggregation and adhesion
ftf	Sucrose-dependent adhesion – synthesis of fructan polymers EPS matrix synthesis – storage of nutrients and hydrating the biofilm
dexA	EPS matrix remodeling – digests glucans for EPS synthesis
fruA	EPS matrix remodeling
vicRKX operon	EPS matrix regulation - Regulation of GTFs glucan synthesys
gcrG	EPS matrix regulation
luxS	EPS matrix regulation - - Regulation of GTFs glucan synthesys Affects biofilm accumulation and structure
rpoA	EPS matrix regulation
ccpA	EPS matrix regulation – up-regulation of EPS synthesis
manL	EPS matrix regulation - – up-regulation of EPS synthesis
brpA	Regulation of biofilm formation Stress tolerance – acidic and oxidative stress Genetic expression regulation
galK	Galactose metabolism
comC	Quorum sensing stimulating precursor
comA	QS – CSP specific secretion apparatus - ATP-binding cassette transporter
comB	QS – CSP specific secretion apparatus – Accessory protein
comD	QS – Histidine Kinase sensor protein Antibiotic resistance
comE	QS – Cognate response regulator Antibiotic resistance
comX	QS – Regulator and inducer of late and delayed gene induction and repression
ciaRH	Survival to low pH and reactive oxygen Antibiotic resistance
vicRKX operon	Survival to low pH and reactive oxygen Antibiotic resistance
liaSR	Survival to low pH and reactive oxygen Antibiotic resistance
ropA	Trigger factor associated with survival to low pH Genetic expression regulation
relA	Survival to low pH and acid tolerance Genetic expression regulation

dnaK	Molecular chaperone induced by heat shock and acid stress
groEL	Molecular chaperone induced by heat shock and acid stress
groES	Molecular chaperone induced by heat shock and acid stress
rex	Redox suppressor that responds to changes in NADH/NAD+ ratio

Acknowledgments

SS holds an individual FCT fellowship (SFRH/BD/90867/2012).

References

Adil, M., Singh, K., Verma, P. K., Khan, A. U. (2014) "Eugenol-induced suppression of biofilme-forming genes in *Streptococcus mutans*: an approach to inhibit biofilms" in **Journal of Global Antimicrobial Resistance**, 4: 286–292

Banas, J. A. (2004) "Virulence properties of *Streptococcus mutans*" in **Frontiers in Bioscience**, 9:1267-1277

Bitoun, J. P., Liao, S., Yao, X., Ahn, SJ., Isoda, R., Nguyen, A. H., et al. (2012) " BrpA is involved in regulation of cell envelope stress responses in *Streptococcus mutans*" in **Applied and Environmental Microbiology**, 78:2914-2922

Bitoun, J. P., Nguyen, A. H., Fan, Y., Burne, R. A., Wen, Z. T. (2011) "Transcriptional repressor Rex in involved in regulation of oxidative stress response and biofilm formation by *Streptococcus mutans*" in **FEMS Microbiology Letters**, 320:110-117

Carlsson, J. (1968) "A numerical taxonomic study of human oral streptococci" in **Odontologisk revy**, 19:137-160

Clarke, J. K. (1924) "On the bacterial factor in the Ætiology of dental caries" in **The British Journal of Experimental Pathology**, 5:141-147

Costa, E. M., Silva, S., Tavaria, F. K., Pintado, M. M. (2013) "Study of the effects of chitosan upon *Streptococcus mutans* adherence and biofilm formation" in **Anaerobe**, 20:27-31

Dong, L., Tong, Z., Linghu, D., Lin, Y., Tao, R., Liu, J., et al. (2012) "Effects of sub-minimum inhibitory concentrations of antimicrobial agents on *Streptococcus mutans* biofilm formation" in **International Journal of Antimicrobial Agents**, 39:390-395

Duque, C., Stipp, R. N., Wang, B., Smith, D. J., Höfling, J. F., Kuramitsu, H. K., et al. (2011) "Downregulation of GbpB, a component of the VicRK regulon, affects biofilm formation and cell surface characteristics of *Streptococcus mutans*" in **Infection and Immunity**, 79:786-796

Klein, M. I., DeBaz, L., Agidi, S., Lee, H., Xie, G., Lin, A. HM., et al. (2010) "Dynamics of *Streptococcus mutans* transcriptome in response to starch and sucrose during biofilm development" in **PLoS ONE**, 5:1-13

Klein, M. I., Duarte, S., Xiao, J., Mitra, S., Foster, T. H., Koo, H. (2009) "Structural and molecular basis of the role of starch and sucrose in *Streptococcus mutans* biofilm development" in **Applied and Environmental Microbiology**, 75:837-841

Klein, M. I., Xiao, J., Lu, B., Delahunty, C. M., Yates J. R., Koo, H. (2012) "*Streptococcus mutans* protein synthesis during mixed-species biofilm development by high-throughput quantitative proteomics" in **PLoS ONE**, 7:1-13

Li, YH., Lau P. C. Y., Lee J. H., Ellen, R. P., Cvitkovitch, D. G. (2001) "Natural genetic transformation of *Streptococcus mutans* growing in biofilms" in **Journal of Bacteriology**, 183:897-908

Li, YH., Tang, N., Aspiras, M. B., Lau, P. C., Lee, J.H., Ellen, R.P., et al. (2002) "A quorum-sensing signaling system essential for genetic competence in *Streptococcus mutans* is involved in biofilm formation" in **Journal of Bacteriology**, 184:2699-2708

Loesche, W. J. (1986) "Role of *Streptococcus mutans* in human dental decay" in **Microbiological Reviews**, 50:353-380

Mattos-Graner, R. O., Napimoga, M. H., Fukushima, K., Duncan, M. J., Smith, D. J. (2004) "Comparative analysis of Gtf isozyme production and diversity isolates of *Streptococcus mutans* with different biofilm growth phenotypes" in **Journal of Clinical Microbiology**, 42:4586-4592

Merritt, J., Qi, F., Goodman, S. D., Anderson, M. H., Shi, W. (2003). "Mutation of *luxS* affects biofilm formation in *Streptococcus mutans*" in **Infection and Immunity**, 71:1972-1979

Nakano, K., Tsuji, M., Nishimura, K., Nomura, R., Ooshima, T. (2006) "Contribution of cell surface protein antigen PAc of *Streptococcus mutans* to bacteremia" in **Microbes and Infection**, 8:114-121

Senadheera, D., Cvitkovitch, D. G. (2008) "Quorum sensing and biofilm formation by *Streptococcus mutans*" in **Bacterial Signal Transduction: Networks and Drug Targets,** edited by Utsumi, R., Springer New York, New York, USA, pp:178-188

Shemesh, M., Tam, A., Feldman, M., Steinberg, D. (2006) "Differential expression profiles of *Streptococcus mutans ftf, gtf* and *vicR* genes in the presence of dietary carbohydrates at early and late exponential growth phases" in **Carbohydrate Research**, 341:2090-2097

Shemesh, M., Tam, A., Steinberg, D. (2007) "Differential gene expression profiling of *Streptococcus mutans* cultured under biofilm and planktonic conditions" in **Microbiology**, 153:1307-1317

Suntharalingam, P., Cvitkovitch, D. G. (2005) "Quorum sensing in streptococcal biofilm formation" in **Trends in Microbiology**, 13:3-6

Suntharalingam, P., Senadheera, M. D., Mair, R. W., Lévesque, C. M., Cvitkovitch, D. G. (2009) "The LiaFSR system regulates the cell envelope stress response in *Streptococcus mutans*" in **Journal of Bacteriology**, 191:2973-2984

Wen, Z. T., Baker, H. V., Burne, R. A. (2006) "Influence of BrpA on critical virulence attributes of *Streptococcus mutans*" in **Journal of Bacteriology**, 188:2983-2992

Wen, Z. T., Burne, R. A. (2002) "Functional genomics approach to identifying genes required for biofilm development by *Streptococcus mutans*" in **Applied and Environmental Microbiology**, 68:1196-1203

Wen, Z. T., Burne, R. A. (2004) "LuxS-mediated signaling in *Streptococcus mutans* is involved in regulation of acid and oxidative stress tolerance and biofilm formation" in **Journal of Bacteriology**, 186:2682-2691

Wen, Z. T., Suntharaligham, P., Cvitkovitch, D. G., Burne, R. A. (2005) "Trigger factor in *Streptococcus mutans* is involved in stress tolerance, competence development, and biofilm formation" in **Infection and Immunity**, 73:219-225

Yoshida, A., Ansai, T., Takehara, T., Kuramitsu, H. K. (2005) "LuxS-based signaling affects *Streptococcus mutans* biofilm formation" in **Applied and Environmental Microbiology**, 71:2372-2380

Yoshida, A., Kuramitsu, H. K. (2002) "Multiple *Streptococcus mutans* genes are involved in biofilme formation" in **Applied and Environmental Microbiology**, 68:6283-6291

Chapter 11

Gardenerella vaginalis gene expression in biofilms

Joana Castro[1,2], Daniela Machado[1], Nuno Cerca[1]*
[1]Centre of Biological Engineering, University of Minho, Campus de Gualtar, 4710-057, Braga, Portugal.
[2]Instituto de Ciências Biomédicas de Abel Salazar, Universidade do Porto, Porto, Portugal
*** nunocerca@ceb.uminho.pt**

Contents of chapter

Introduction to the chapter

Bacterial vaginosis (BV) is the leading vaginal disorder worldwide and it has been associated with important public health problems. Due to their complexity and to the diversity of microorganisms involved, BV etiology remains elusive. Recent studies reported that BV is related with the presence of biofilm composed mainly by *Gardnerella vaginalis*. In the last decade, the novel data about vaginal transcriptome has been provided new insights regarding BV-associated bacteria and its interaction with immune system. Consequently, transcriptomic studies are helpful in knowledge of BV pathogenesis and their findings have been applied to fight this disorder more effectively.

Healthy vaginal microbiota

Microbial communities have a high influence in human health and quality of life. Particularly, the bacterial community of the human vagina has a profound impact on the health of women, since inhabitant microorganisms determine the biochemical and inflammatory profile of the vaginal environment (Vitali et al, 2007). In premenopausal women, the healthy vaginal microbiota is typically dominated by lactobacilli, being the most common species *Lactobacillus crispatus*, *L. iners*, *L. gasseri* and *L. jensenii*. Additionally, a great diversity of other microorganisms including *Gardnerella vaginalis*, *Staphylococcus aureus*, *Escherichia coli*, *Candida albicans* and *Prevotella* species, can be present, but in lower number (Verhelst et al, 2005). In the human vagina, lactobacilli play a protective role, maintaining of healthy vaginal ecosystem and preventing urogenital infections. Basically, lactobacilli inhibit the overgrowth of pathogens and opportunistic microorganisms through its capacity to adhere and compete for adhesion sites in the vaginal epithelium (Castro et al, 2013; Machado et al, 2013a; Machado et al, 2013b) and production of antimicrobial compounds including hydrogen peroxide, lactic acid and bacteriocins (Aroutcheva et al, 2001a).

Etiopathogenesis of BV

Despite of the protective role of lactobacilli in vagina, disturbances in the vagina microflora can occur, being BV the most common vaginal disturbance in women of reproductive age. It has been linked to serious public health consequences including preterm birth, postoperative infections and the acquisition and transmission of several sexually transmitted agents (reviewed in Livengood, 2009). BV is characterized by a complex imbalance of vaginal microflora which includes the loss of lactobacilli, mainly hydrogen peroxide and lactic acid producing strains, and concurrent overgrowth of anaerobic bacteria, such as *G. vaginalis*, *Atopobium vaginae*, *Mycoplasma hominis*, *Ureaplasma parvum*, *Bacteroides* spp., *Mobiluncus* spp. and *Prevotella* spp. (Verhelst et al, 2005).

Currently, the cause of BV remains controversial and there are two main hypotheses that try to explain its development. The first is the polymicrobial hypothesis, which infers that BV is caused by a mixture of pathogenic bacteria, principally anaerobes. The second is that a single pathogenic species, in many cases *G. vaginalis*, is the etiological agent of BV (reviewed in Srinivasan and Fredricks, 2008).

Recently, we showed that while one BV-associated *G. vaginalis* strain was able to displace a protective layer of vaginal lactobacilli and colonize HeLa epithelium cells, this did not occurred with a non-BV strain (Castro et al, 2013). By displacing lactobacilli, adhered *G. vaginalis* will then start to form a biofilm that will allow later incorporation of secondary microorganisms and this polymicrobial biofilm will become recalcitrant to antimicrobial therapy (Swidsinski et al, 2008).

G. vaginalis as a pathogen

G. vaginalis is a facultative anaerobic bacterium, but some strict anaerobic strains exist. This bacterium belongs to the *Bifidobacteriaceae* family and is often described as a Gram-variable organism but it has a Gram-positive wall type (reviewed in Catlin, 1992). *G. vaginalis* is considering a fastidious organism and requires complex medium for growth. Some different selective media have been used to isolate *G. vaginalis* from clinical specimens, such as columbia blood agar supplemented with horse blood (Cereija et al, 2013) and human blood-bilayer-tween agar (reviewed in Catlin, 1992).

Although *G. vaginalis* has been identified as one of the frequent and predominant colonizer of the vaginal tract of women diagnosed with BV, it is also commonly isolated from vaginal microbiota of healthy women (Aroutcheva et al, 2001b; Harwich et al, 2010). Consequently, there has been vigorous debate in the literature concerning the role of *G. vaginalis* in the etiology of BV (reviewed in Schwebke et al, 2014). Nevertheless, other bacterial species are also commonly found in BV women; however their role in BV is not clear. Interestingly, our research group demonstrated that *G. vaginalis* had significant higher virulence potential than other BV-associated species (Alves et al, 2014).

Taking in consideration previous remarks, it was hypothesized that colonization by a subset of *G. vaginalis* is the trigger for BV development (Harwich et al, 2010; Yeoman et al, 2010). Consequently, studies of *G. vaginalis* is of significant interest to both clinicians and researchers.

Clinical features and the epidemiology of BV

Clinically, the profuse vaginal discharge and fishy vaginal odour are the two most commonly reported symptoms of BV. However, some women with BV do not report symptoms and it is unknown whether asymptomatic BV resolves, remains stable, or becomes symptomatic (Koumans et al, 2007). Moreover, this disorder is generally diagnosed by the interpretation of symptoms and signs of patient (Amsel criteria) (Amsel et al, 1983) or microscopic examination of Gram-stained vaginal smear (Nugent criteria) (Nugent et al, 1991). However, these diagnostic tools are subjective and, therefore, alternative methods for BV diagnosis have been investigated (reviewed in Africa, 2013). Recently, major advances in the BV diagnosis have been achieved using molecular techniques. The molecular methodologies, such as PCR (Henriques et al, 2012a), fluorescence *in situ* hybridization (Machado et al, 2013c) and other techniques described in the last section of this chapter, have allowed detecting and identifying the main BV-associated bacteria and also they have improved our understanding of how microbes interact with each other and with immune system.

Despite the high clinical importance of BV, its true prevalence is unknown since it varies according to the characteristics of studied population, in terms of age, sexual behavior, geographic localization, ethnicity, hygienic practices and socioeconomic status (reviewed in Kenyon et al, 2013). Nevertheless, epidemiological studies indicated that the risk of BV is increased in sexually active women, mainly in women who report sex with other women, and those who report new or a higher number of male sex partners (Fethers et al, 2009; Schwebke and Desmond, 2005). Other risk factors include black ethnicity, poverty, low educational level and douching practice (Koumans et al, 2007). In Portugal, the epidemiological data about prevalence and risk factors of BV are almost non-existent, limiting only to three studies. The first epidemiological study about BV was developed by Guerreiro et al (1998) that reported a prevalence rate of BV of 7% among contraceptive users and established a positive association between use of intra-uterine device and BV occurrence. Later, Henriques et al (2012b) assessed Portuguese doctors' perception of BV prevalence through anonymous questionnaire. They verified that most doctors considered BV a frequent condition in Portugal (74%) with a lower prevalence during pregnancy (55%). Recently,

Silva et al (2014) determined the prevalence of BV (20%) in Portuguese women and correlated the presence of *G. vaginalis* with risk factors for BV. These researchers demonstrated that women with BV have a higher risk of harboring *G. vaginalis*, being *G. vaginalis* colonization linked with increased risk of present itching and odour (Silva et al, 2014).

Due to the high impact of BV in women health, there is a special interest in developing alternative therapies against BV that might be safer and more efficient than antibiotics. This is mainly related with the very high recurrence levels detected in BV, after antimicrobial theraphy (Bradshaw et al, 2006). According to Centers for Disease Control and Prevention (2010), BV treatment involves antibiotics, including metronidazole, clindamycin or tinidazole. Several therapeutic alternatives, such as probiotics, acidification procedure and natural compounds, have been explored in modality of the standalone therapy or adjuvant of antibiotics (reviewed in Donders et al, 2014).

G. vaginalis transcriptomic studies

For more than a half century the virulence mechanism of *G. vaginalis*, identified as one of the key players in the development of BV, has remained poorly understood (Gardner and Dukes, 1955). Recent findings on the comparative genomic analysis of several *G. vaginalis* strains isolated from BV patients and healthy individuals have provided new outcomes on pathogenic potential and experimentally proven virulence features of *G. vaginalis*, namely, bacterial adhesion to epithelium, cytotoxicity and biofilm formation (Harwich et al, 2010; Yeoman et al, 2010).

Recently, virulence studies showed that BV-associated *G. vaginalis* isolates are able to adhere, and thus establish an infection much more readily than non-BV associated isolates (Castro et al, 2013; Harwich et al, 2010). Interestingly, the cytotoxicity activity of BV-associated *G. vaginalis* isolates was also significantly stronger than non-BV isolates, suggesting that differing levels of cytotoxicity may be related to the propensity of a strain to cause disease. This cytotoxicity effect could be due to a pore-forming toxin, belonging a cholesterol-dependent cytolysin family produced by *G. vaginalis*, named vaginolysin (encoded by *vly*), which is able to induce cell dead (Gelber et al, 2008). To determine if differences in expression of *vly* could account for the reduced cytotoxicity associated with the non-BV associated strains, we recently analysed the *vly* expression by quantitative PCR (qPCR). Interestingly, data revealed that on average, BV-associated *G. vaginalis* isolates expressed 2-fold more *vly* than non-BV isolates. However, strain to strain variability found suggests that *vly* expression is not exclusive of BV-associated *G. vaginalis* (personal communication).

Similar cytolysins, like hemolysins, appear to contribute to virulence by making cellular contents more available as a substrate for bacterial growth (Yeoman et al, 2010). Also, sialidase could be increasing the cytotoxic ability of *G. vaginalis*, once it is responsible by degrading several key mucosal protective factors, such as mucins, as well as contributing to exfoliation and detachment of vaginal epithelial cells (Cauci et al, 2008). Nevertheless, recent data did not reveal a direct relationship between hemolysin (*hemo*) or sialidase (*sial*) expression and cytotoxicity, showing that transcriptional expression of the *hemo* and *sial* genes was equivalent in the two groups of strains (personal communication).

The ability of *G. vaginalis* to form a biofilm is marker of potential virulence and cause of recurrent infections (Swidsinski et al, 2008). In 2010, Harwich et al identified a biofilm associated (BAP) family gene, in which called BAP-Like (*bapL*). BAP proteins are large, cell wall-anchored adhesins that can mediate both adherence to host cells and intercellular adherence, thus contributing to biofilm formation. They hypothesized that BapL would be localized to the surface of *G. vaginalis* and would be expressed in biofilm cultures. Using qPCR, *bapL* expression was significantly elevated under conditions that promote the biofilm phenotype relative to planktonic cultures (Harwich et al, 2010). In the track of determining the existence on virulent and avirulent *G. vaginalis*, they investigated whether the disparity in the sequences and expression of *bapL* could translate into differences in biofilm formation activity. Interestingly, the BAP domain sequence between the BV and non-BV strains were quite disparate. Their study showed that a BV-associated strain tend to produce more biofilm growth than a non-BV strain, which would reasonably be consistent with greater virulence (Harwich et al, 2010). Nevertheless, a larger array of BV and non-BV *G. vaginalis* isolates would help confirming the results and understanding strain variations.

As described above, *G. vaginalis* strains have the ability to form biofilms in vagina. However, *G. vaginalis* colonization can ascend to upper genital tract, confirming that the pathogenic effects of BV are not confined to the lower genital tract (Swidsinski et al, 2013). This issue is of high importance in pregnancy, because the fetal membranes are likely the first tissue colonized by an ascending infection, causing an inflammation, and, consequently, the risk of preterm birth. The way in which an infection can lead to inflammation and preterm birth is thought to be mediated by innate immune pattern recognition receptors, namely the Toll-like receptors (TLRs). TLRs are present on several types of cells, and their recognition of a wide variety of pathogens, including bacteria, result in the induction of an antimicrobial and inflammatory cytokine response for pathogen elimination. A recent study analysed the premature rupture of fetal membranes by an induced infection of *G. vaginalis*, as well as other BV-associated bacteria (Abrahams et al, 2013). However, no effect in expression of TLR was detected for *G. vaginalis* by qPCR. Nevertheless, *Mycoplasma hominis* and *Ureaplasma parvum*, commonly present in BV vaginal flora, showed an increased expression of TLRs and consequently pregnancy complications (Abrahams et al, 2013).

Potential of transcriptomics to analyse the vaginal microbiome

In effort to overcome the limitations of traditional BV diagnosis, molecular methodologies have been successfully developed to characterize microbial species, increasing the knowledge considerably with regard to the microbes associated with BV. This can help clinicians to detect uncultivable bacteria that can be involved in the occurrence of BV as well as failure of its treatment.

Recently, Balashov et al (2014) analysed the depletion of *Lactobacillus* species and its association with BV using a multiplex qPCR for the quantitative assessment of the four major lactobacilli: *L. crispatus*, *L. jensenii*, *L. gasseri* and *L. iners*. They showed that *L. crispatus*, *L. jensenii* and to a lesser extent *L. gasseri*, are common in the vagina of healthy women, whereas *L. iners* dominance is associated with BV. Interestingly, studies employing qPCR of vaginal microbiota coupled with Nugent scoring has been used to optimize molecular methods thereby creating thresholds for BV diagnosis with very good sensitivity (95%) and specificity (99%).

To understand the vaginal microbiota in more detail, microarray-based identification can be used after microscopic scoring. Microarrays have the potential to simultaneously assess most of the microorganisms of a complex microbial community, being used to diagnose BV. Remarkably, Dols et al (2011) noted the coexistence of *Leptotrichia* and *Prevotella* species with *G. vaginalis*. However, studies are needed to examine whether these microorganisms have some sort of codependency, or increased virulence. An advantage of microarrays, compared to qPCR, is the possibility to obtain new data on coexistence of a very large number of pathogens.

By employing a full 16S pyrosequencing strategy and developing a classification method of short 16S pyrosequencing reads, Shipitsyna et al (2013) studied the composition of the vaginal microbiota in women of reproductive age. The depletion of *Lactobacillus* species combined with the presence of either *G. vaginalis* or *A. vaginae* was a highly accurate BV predictor.

Advanced molecular techniques, such as RNA sequencing, have been used to uncover genes and pathways that potentially differentiate healthy vaginal microbial communities from those in the dysbiotic state of BV. Macklaim et al (2013) showed that *L. iners* was present in a healthy and BV flora but showing a differing expression profile in BV compared to healthy. Notably, in a BV environment *L. iners* increased expression of a cholesterol-dependent cytolysin, a mucin, and a related metabolic enzymes. Studies like these highlights the importance of understanding the functional activity of the bacterial flora in addition to characterizing the population structure when investigating the human microbiome.

Overall, examining the vaginal transcriptome has highlighted the complexity of this environment. The large difference in bacterial function between BV and healthy conditions emphasizes our need to better understand the interactions between species in the context of the rest of the community and the host. Nevertheless, molecular methodologies and its implementation in the laboratory for BV diagnosis are time-consuming, costly and also involve rigorous optimization. However, all these requirements are being overcome and further implementation of these methods will be suitable in the future in all clinical microbiology laboratories.

Acknowledgments

JC and DM hold an individual FCT fellowship (SFRH/BD/93963/2013 and SFRH/BD/87569/2012 respectivelly).

References

Abrahams,V., J. Potter, et al. (2013) "Bacterial modulation of Human fetal membrane toll-like receptor expression." in **American Journal of Reproductive Immunology**, 69: 33–40

Africa, C. (2013) "Efficacy of methods used for the diagnosis of bacterial vaginosis." in **Expert Opinion on Medical Diagnostics**, 7: 189–200

Alves, P., J. Castro, et al. (2014) "*Gardnerella vaginalis* outcompetes 29 other bacterial species isolated from BV patients in an *in vitro* biofilm formation model." in **Journal of Infectious Diseases**, 210: 593–596

Amsel, R., P. Totten, et al. (1983) "Non-specific vaginitis: diagnostic techniques and microbial and epidemiologic associations." in **American Journal of Medicine**, 74: 14–22

Aroutcheva, A., D. Gariti, et al. (2001a) "Defense factors of vaginal lactobacilli." in **American Journal of Obstetrics and Gynecology**, 185: 375–379

Aroutcheva, A., J. Simoes, et al. (2001b) "*Gardnerella vaginalis* isolated from patients with bacterial vaginosis and from patients with healthy vaginal ecosystems." in **Clinical Infectious Diseases**, 33: 1022–1027

Balashov, S., E. Mordechai, et al. (2014) "Multiplex quantitative polymerase chain reaction assay for the identification and quantitation of major vaginal lactobacilli." in **Diagnostic Microbiology and Infectious Disease**, 78: 321–327

Bradshaw, C., A. Morton, et al. (2006) "High recurrence rates of bacterial vaginosis over the course of 12 months after oral metronidazole therapy and factors associated with recurrence." in **Journal ofInfectious Diseases**, 193: 1478–1486

Castro, J., A. Henriques, et al. (2013) "Reciprocal interference between *Lactobacillus* spp. and *Gardnerella vaginalis* on initial adherence to epithelial cells." in **International Journal of Medical Sciences**, 10: 1193–1198

Catlin, B. (1992) "*Gardnerella vaginalis*: Characteristics, clinical considerations, and controversies." in **Clinical Microbiology Reviews**, 5: 213–237

Cauci, S., J. Culhane, et al. (2008) "Among pregnant women with bacterial vaginosis, the hydrolytic enzymes sialidase and prolidase are positively associated with interleukin-1beta." in **American Journal of Obstetrics and Gynecology**,198: 132.e1–7

Centers for Disease Control and Prevention (2010) "Sexually transmitted diseases treatment guidelines." in **Morbidity and Mortality Weekly Report**, 59: 56–62

Cereija, T., J. Castro, et al. (2013) "Influence of anaerobic conditions on vaginal microbiome recovery from bacterial vaginosis patients." in **Sexually Transmitted Infections**, 89: 307

Dols, J., P. Smit, et al. (2011) "Microarray-based identification of clinically relevant vaginal bacteria in relation to bacterial vaginosis." in **American Journal of Obstetrics and Gynecology**, 204: 305.e1–7

Donders, G., J. Zodzika, et al. (2014) "Treatment of bacterial vaginosis: what we have and what we miss." in **Expert Opinion on Pharmacotherapy**, 15: 645–657

Fethers, K., C. Fairley, et al. (2009) "Early sexual experiences and risk factors for bacterial vaginosis." In **Journal of Infectious Diseases**, 200: 1662–1670

Gardner, H. and C. Dukes (1955) "*Haemophilus vaginalis* vaginitis. A newly defined specific infection previously classified 'nonspecific vaginitis'." in **American Journal of Obstetrics and Gynecology**, 69:962–976

Gelber, S., J. Aguilar, et al. (2008) "Functional and phylogenetic characterization of vaginolysin, the human-specific cytolysin from *Gardnerella vaginalis*." in **Journal of Bacteriology**, 190: 3896–3903

Guerreiro, D., M. Gigante, et al. (1998) "Sexually transmitted diseases and reproductive tract infections among contraceptive users." in **International Journal of Gynecology and Obstetrics**, 63: Suppl167–173

Harwich, M., J. Alves, et al. (2010) "Drawing the line between commensal and pathogenic *Gardnerella vaginalis* through genome analysis and virulence studies." in **BMC Genomics**, 11: 375

Henriques, A., T. Cereija, et al. (2012a) "*In silico* vs *in vitro* analysis of primer specificity for the detection of *Gardnerella vaginalis, Atopobium vaginae* and *Lactobacillus* spp." in **BMC Research Notes**, 5:637

Henriques, A., J. Martinez-de-Oliveira, et al. (2012b) "Doctor's perception on bacterial vaginosis in Portugal: prevalence, diagnostic methods and choice of treatment." in **Sexually Transmitted Diseases,** 88:421

Kenyon, C., R. Colebunders et al. (2013) "The global epidemiology of bacterial vaginosis: a systematic review." in **American Journal of Obstetrics and Gynecology** 209: 505–523

Koumans, E., M. Sternberg, et al. (2007) "The prevalence of bacterial vaginosis in the United States, 2001-2004; associations with symptoms, sexual behaviors, and reproductive health." in **Sexually Transmitted Diseases**, 34:864–869

Livengood, C. (2009) "Bacterial vaginosis: an overview for 2009." in **Review in Obstetrics and Gynecology**, 2: 28–37

Machado, A., D. Salgueiro, et al. (2013a) "Quantitative analysis of initial adhesion of bacterial vaginosis associated anaerobes to ME-180 cells." In **Anaerobe**, 23:1–4

Machado, A., K. Jefferson, et al. (2013b) "Interactions between *Lactobacillus crispatus* and BV-associated bacterial species in initial attachment and biofilm formation." in **International Journal of Molecular Sciences**, 14: 12004–12012

Machado, A., C. Almeida, et al. (2013c) "Fluorescence *in situ* Hybridization method using Peptide Nucleic Acid probes for rapid detection of *Lactobacillus* and *Gardnerella* spp." in **BMC Microbiology**, 13:82

Macklaim, J., A. Fernandes, et al. (2013) "Comparative meta-RNA-seq of the vaginal microbiota and differential expression by *Lactobacillus iners* in health and dysbiosis." in **Microbiome**, 1:12

Nugent, R., M. Krohn, et al. (1991) "Reliability of diagnosing bacterial vaginosis is improved by a standardized method of Gram stain interpretation." in **Journal of Clinical Microbiology**, 29:297–301

Schwebke, J. and R. Desmond, (2005). "Risk factors for bacterial vaginosis in women at high risk for sexually transmitted diseases." in **Sexually Transmitted Diseases**, 32: 654–658

Schwebke, J., C. Muzny, et al. (2014) "Role of *Gardnerella vaginalis* in the pathogenesis of bacterial vaginosis: a conceptual model." in **Journal of Infectious Diseases**, 210: 338–343

Shipitsyna, E., A. Roos, et al. (2013) "Composition of the vaginal microbiota in women of reproductive age-sensitive and specific molecular diagnosis of bacterial vaginosis is possible?" in **Plos One**, 8:e60670

Silva, D., A. Henriques, et al. (2014) "Prevalence of *Gardnerella vaginalis* and *Atopobium vaginae* in Portuguese women and association with risk factors for bacterial vaginosis." in **International Journal of Gynecology and Obstetrics**, 124:178–179

Srinivasan, S., D. Fredricks (2008) "The human vaginal bacterial biota and bacterial vaginosis." in **Interdisciplinary Perspectives on Infectious Diseases**, 2008:750479

Swidsinski, A., W. Mendling, et al. (2008) "An adherent *Gardnerella vaginalis* biofilm persists on the vaginal epithelium after standard therapy with oral metronidazole." in **American Journal Obstetrics and Gynecology**, 198: 97e1–6

Swidsinski, A., H. Verstraelen, et al. (2013) "Presence of a polymicrobial endometrial biofilm in patients with bacterial vaginosis." in **Plos One**, 8: e53997

Verhelst, R., H. Verstraelen, et al. (2005). "Comparison between Gram stain and culture for the characterization of vaginal microflora: definition of a distinct grade that resembles grade I microflora and revised categorization of grade I microflora." in **BMC Microbiology**, 5: 61

Vitali, B., C. Pugliese et al. (2007) "Dynamics of vaginal bacterial communities in women developing bacterial vaginosis, candidiasis, or no infection, analyzed by PCR-denaturing gradient gel electrophoresis and real-time PCR." in **Applied and Environmental Microbiology**, 73: 5731–5741

Yeoman, C., S. Yildirim, et al. (2010) "Comparative genomics of *Gardnerella vaginalis* strains reveals substantial differences in metabolic and virulence potential." in **Plos One** 5: e12411

Chapter 12

Bacillus cereus a foodborne pathogen involved in biofilm infections

Anabela Borges, Joana Malheiro, Manuel Simões*
LEPABE, Department of Chemical Engineering, Faculty of Engineering, University of Porto, Rua Dr. Roberto Frias, s/n, 4200-465 Porto, Portugal
*** mvs@fe.up**

Contents of chapter

Introduction to the chapter

Food provides the ideal environment for the development of microorganisms. Hence, the contamination of food contact surfaces and equipments with pathogenic microorganisms is almost inevitable. Although cleaning and disinfection procedures are regularly applied, these are not sufficient, and the pathogens are frequently transmitted to food. Diseases caused by the ingestion of contaminated foods has economic and public health impact worldwide (reviwed in Newell, 2010; McLinden, 2014). Many pathogenic opportunist microorganisms such as *Listeria monocytogenes, Staphylococcus aureus, Clostridium perfringens, Bacillus cereus, Escherichia coli, Campylobacter jejuni, Salmonella* sp. and *Pseudomonas aeruginosa* have been reported as the causal agents of foodborne diseases (McCabe-Sellers, 2004; Behling, 2010; reviwed in Newell, 2010; Choi, 2013).

B. cereus is among the most relevant Gram-positive foodborne pathogens, although it is frequently devalued and underreported due to their widespread distribution in nature (reviwed in Granum, 1997; Lee, 2010). It easily adheres to food processing surfaces and forms a biofilm, which due to their persistence impairs the quality and safety of food products, constituting a vehicle for human infection (Oosthuizen, 2002; Wijman, 2007; Nam, 2014). Apart from the involvement of *B. cereus* biofilms in food spoilage and food poisonings, these microbial communities have also an important role in hospital environments (reviwed in Bottone, 2010). Nosocomial infections by *B. cereus* are normally associated with invasive procedures, bodily trauma or skin lesions that provide a portal of entry to the body. The use of "foreign bodies" or indwelling medical devices such as, contact lens, intravascular devices, prosthetic valves, pacemakers, breast implants and urethral catheters, are also a predispositions factors for acquire

B. cereus infections (reviwed in Drobniewski, 1993). In some situations, the systemic infections are related with the acquisition of *B. cereus* from exogenous source, like food or water. This may result from gastric invasion and mucosal necrosis that leads to the spread of the bacterium to other sites (Funada, 1988; reviwed in Bottone, 2010). The colonization of medical devices and subsequent formation of biofilms in these surfaces increases the severity of infections. Consequently, due to the properties of biofilm cells, the infections persist even after device removal (reviwed in Bottone, 2010). Bacteria within biofilms experience many alterations in gene regulation that lead biofilm cells to become different from their planktonic counterparts. In fact, the cells in biofilms exhibit unique patterns of gene expression that contribute to the increase of the resistance.

This chapter provides an overview about *B. cereus* group and their involvement in foodborne diseases. In addition, the impact of *B. cereus* in industrial and hospital environments will be focused. The concomitant role of sporulation and biofilm formation in the relationship between persistence of *B. cereus* and infections will be also highlighted. As the profile of genes expressed in biofilms is different from the planktonic state, a summary about some of the genes associated with *B. cereus* biofilm formation will be presented.

Bacillus cereus group

Members of the *Bacillus* genus, are widely distributed in the environment although the primary habitat is the soil. Their natural niche range from vegetation to insect intestine, dairy products, food and some species are pathogenic to animals and humans (reviwed in Rasko, 2005; Stenfors, 2008). The microorganisms of the *Bacillus* genus are Gram-positive, aerobic or facultative anaerobic, motile and rod-shaped spore-forming (reviwed in Drobniewski, 1993; Kotiranta, 2000). Although their optimal temperature growth is observed in the range of 25ºC to 37ºC, certain species are psychrophilic and other are thermophilic, which make this genus highly heterogeneous. The G+C content of the DNA can vary from 32% to 69% between different species (reviwed in Drobniewski, 1993). Most strains are catalase positive, possess peritrichous flagella, and sporulate in air, characteristics that help in the differentiation of members of this genus from the clostridia (reviwed in Drobniewski, 1993). The *Bacillus* genus is large, including more than 70 species with a great genetic diversity that can be divided into five (based on 16S rRNA phylogeny) or six (based on phenotypic features) groups (Økstad, 2011).

The *B. cereus* group of organisms, a subgroup within the *Bacillus* genus, also known as *B. cereus sensu lato,* comprises six approved species, namely *Bacillus anthracis, B. cereus (sensu stricto), Bacillus thuringiensis, Bacillus mycoides, Bacillus pseudomycoides* and *Bacillus weihenstephanensis* (reviwed in Drobniewski, 1993). However, only some of them are of human interest (*B. anthracis, B. cereus* and *B. thuringiensis*) due to their known pathogenic potential. As has been observed, all the members are closely related and some share a very similar chromosomal content. It is therefore suggested that the entire group represents a single species, *B. cereus sensu lato* (reviwed in Jensen, 2003; Rasko, 2005).

B. anthracis is the etiological agent of anthrax, an acute fatal disease found primarily among herbivores, but all mammals are susceptible and isolated cases of human infections has been observed (reviwed in Jensen, 2003; Rasko, 2005). *B. thuringiensis* has long been regarded as an insect pathogen and can be commercially used as biological insecticide in crop protection. *B. cereus* is an opportunistic pathogen, commonly found in a wide range of environments (e.g. soils, clays, sediments, dust, mineral water and plants) and foods (e.g. rice, milk/dairy products, spices, dried foods, raw/processed foods and vegetables), with the proficiency to cause food poisoning and serious infections in mammals (reviwed in Granum, 1997; Schoeni, 2005; Stenfors, 2008).

Bacillus cereus and foodborne diseases

In recent years, *B. cereus* has been reported as responsible for a growing number of foodborne diseases in developed countries (reviwed in Kotiranta, 2000). Episodes of *B. cereus* food poisoning results from ingestion of contaminated food (normally heat treated foods), sometimes due to improper storage after cooking. It can cause two distinct types of food illness, characterized either by diarrhea and abdominal distress (diarrhoeal form) or by nausea and vomiting (emetic form) (reviwed in Granum, 1997). These gastrointestinal diseases are due to the synthesis of specific toxins by *B. cereus*. The emetic syndrome has often been associated with the eating of carbohydrate-rich foods like rice and pasta and is caused by a heat-stable emetic toxin, the cereulide, which is preformed in food before ingestion (reviwed in Stenfors, 2008). The diarrhoeal type is due to enterotoxin production (haemolysin BL – *hbl*, nonhaemolytic enterotoxin – *nhe* and cytotoxin K – *cytK*) during the vegetative growth of *B. cereus* in the small intestine, and is mainly linked with consumption of meat, vegetables and milk products (reviwed in Drobniewski, 1993; Schoeni, 2005; Stenfors, 2008). Both types of food poisoning are underreported, because they are generally mild and self-limiting and usually do not last for more than 24 hours (reviwed in Granum, 1997). Moreover, it requires little supportive therapy and no antimicrobial treatment. However, more serious and even lethal cases of bloody and emetic poisoning have been occasionally described (reviwed in Ramarao, 2013).

Bacillus cereus pathogenicity

In addition to the familiar role of *B. cereus* in foodborne illnesses and intoxications, when the access to mammalian tissues is allowed, it behaves as an opportunistic pathogen capable of causing a wide variety of infections (Økstad, 2011). This organism is not only implicated in gastrointestinal pain, but it is also associated with severe extra-intestinal disorders, ranging from systemic (bacteremia, septicemia, endocarditis, meningitis, gastritis, osteomyelitis, pneumonia and urinary tract infections - UTIs) to localized (cutaneous/oral infection and endophthalmitis) infections (reviwed in Drobniewski, 1993; Kotiranta, 2000). These nongastrointestinal infections can occur in healthy individuals but principally in individuals with particular underlying conditions such as, immunologically compromised patients, neonates, intravenous drug abusers, patients recovering from surgery or with traumatic wounds/indwelling catheters (reviwed in Drobniewski, 1993; Rasko, 2005). *B. cereus* systemic infections imposes substantial morbidity and mortality of infected individuals.

The pathogenicity and virulence of *B. cereus* in non-intestinal infections depend of its ability to colonize and persist in the host and subsequently to invade/destroy the tissues. Besides, the production of various toxins and enzymes is also of great importance. These secreted compounds include, in addition to the referred pore-forming enterotoxins and emesis-inducing toxin, necrotizing hemolysins, proteases and phospholipases (reviwed in Drobniewski, 1993; Bottone, 2010). Normally, the production of enterotoxins and emetic toxins are linked to the gastrointestinal syndromes and the last ones to the nongastrointestinal diseases. The production of beta-lactamases, which is enzymes that degrade the β-lactamic antibiotics, are also other aspect related to *B. cereus* pathogenicity (reviwed in Kotiranta, 2000).

The role of spores and biofilms in *Bacillus cereus* infections

The nonfastidious nature and ubiquitous presence of *B. cereus*, in combination with the ability to form spores, improves its spreading capacity, even in hostile environments (reviwed in Kotiranta, 2000; Stenfors, 2008). Indeed, the production of spores is a prominent feature of *B. cereus*. Spores are much more resistant than vegetative cells to heat/chemical treatment and other refractory conditions, such as, freezing, drying and radiation, resulting in *B. cereus* survival for long time periods (Nam, 2014). In the industrial settings, *B. cereus* is found as a contaminant of many food products, especially dairy products, causing their spoilage (Pirttijärvi, 2000). Undeniably, contamination of food production systems with *B. cereus* is almost inevitable, because of the hydrophobic nature of spores. This feature allows them to adhere to food-contact surfaces, persist in equipment's and easily spread to all kinds of food (Wijman, 2007; reviwed in Bottone, 2010). In this sense, *B. cereus* causes numerous problems in the quality of the end-product and consequently food poisonings, due to ingestion of contaminated products.

The ubiquitous nature and higher resistance to disinfection of microorganisms that form spores, like *B. cereus*, facilitates also their distribution in hospital environments, where they can constitute a reservoir for diseases. For that reason, this microorganism is frequently involved in nosocomial infections (reviwed in Drobniewski, 1993). The identified sources of contamination are usually associated to the presence of dusts, air or water containing vegetative cells or spores and include, ventilation/bronchoscopy equipments, linen, gloves, gauzes, hands of healthcare workers, alcohol-based hand wash solution, sample collection tubes, balloons used in manual ventilation, reused towels and intravenous catheters (reviwed in Bottone, 2010). When contamination occurs, the high hydrophobicity, low surface charge and the presence of appendages in the *B. cereus* spores are involved in their adhesion to different tissues/organs like epithelial cells of the small intestine or cells of respiratory mucosa (Andersson, 1998). Effectively, these characteristics are important during infection (Andersson, 1998). Moreover, when the spores germinate, they release high amounts of tissue-destructive toxins (reviwed in Bottone, 2010). There is no doubt that sporulation in *B. cereus* appears to be not only a way of survival, but also a strategy for its pathogenesis.

It is known that *B. cereus* is capable to form biofilm in different settings and on a variety of surfaces. In addition to their involvement in the impairment of food quality, biofilms of *B. cereus* represent also a significant health hazard (Oosthuizen, 2002; Wijman, 2007; Simões, 2008). In fact, biofilm formation is required for the permanence of *B. cereus* strains in various environments, namely host intestinal tract, food matrices and healthcare settings (Kamar, 2013). For example, in hospital environment, the formation of biofilms play an important role in the adhesion of *B. cereus* to catheters or other indwelling medical devices (Kuroki, 2009). The attached cells are difficult to eliminate with in use therapy resulting frequently in recurrent and chronic infections. Moreover, this sessile mode of growth increases the resistance of *B. cereus* to disinfectants, enabling thus their persistence in hospital facilities (Ryu, 2005; Kuroki, 2009). Biofilm formation in this microorganism contributes also for the survival of spores, providing them a protected environment (Wijman, 2007). Hence, the capacity of *B. cereus* cells to form complex communities, is another factor that may contribute to the pathogenicity and virulence of this bacterium (Auger, 2009). In this context, research about biofilm formation with this organism are important for economic and health-related purposes.

Gene expression profile of *Bacillus cereus* linked to biofilm formation

The formation of a biofilm leads to the expression of a different transcriptome and consequently a different proteome and phenotype, when compared with planktonic bacteria. The main genes differentially expressed in biofilm formation of Gram-positive bacteria, such as *Bacillus* spp., are involved in motility and chemotaxis, membrane bioenergetics and sugar catabolism (reviwed in Sauer, 2003). In addition, the genes associated with sporulation, transport and metabolism are highly expressed during the formation of *B. cereus* biofilm (Table 12.1). For example, *ptsI* might be one of the most important genes involved in biofilm formation and colonization. In addition, Δ*ptsI* mutant strain has a reduced capacity to grow on a medium with glucose as carbon source (Xu, 2014). Furthermore, the peptidase coded by *cwpFM* may be involved in bacteria cell wall dynamics, influencing biofilm formation, motility and virulence (Tran, 2010).

Table 12.1 | Function of several genes involved in *B. cereus* biofilm formation

Gene	Function	Influence in biofilm formation
ptsI (Xu, 2014)	Enzyme I of the phosphotransferase system responsible for the regulation of carbohydrate metabolism	Gene deletion reduces the capacity to grow on a medium with glucose as carbon source and so influences biofilm formation
cwpFM (Tran, 2010)	Peptidase involved in adhesion and motility	Gene absence reduces biofilm formation by interfering with adhesion and motility
fla (Houry, 2010)	Flagellar apparatus responsible for bacteria adhesion and motility	Gene absence reduces biofilm formation excepting for flow systems
plcR (Gohar, 2002)	Regulator of virulence factors secretion	Gene absence reduces biofilm pathogenicity by diminishing the production of flagellins
codY (Hsueh, 2008; Fagerlund, 2014)	Repressor of branched amino-acids biosynthesis	Gene absence reduces biofilm formation possibly by the repression of an 130 KDa extracellular protease
spoOA/σ^H (Liaqat, 2013; Fagerlund, 2014)	Activates and triggers asymmetric sporulation division and transcription of spoIIA, spoIIE and spoIIG/ Sigma factor that controls expression of proteins involved in the initial stages of sporulation	Gene absence inhibits biofilm formation by interfering with matrix production and sporulation

Motility is one of the first mechanisms that influence *B. cereus* biofilm formation. Besides, motility is necessary to reach the surface and the air-liquid interface, where biofilm is suitable to form, and is also important to promote the recruitment of planktonic bacteria (Houry, 2010). *B. cereus* can move over surfaces by two types of motility, swarming and sliding, depending on nutrient availability. Swarming motility (flagellum-driven motility) in *B. cereus* can play an important role in host pathogen interaction, because it leds to increase secretion of specific virulence factors, like as haemolysin BL (reviwed in Senesi, 2010).

Effectively, toxic effects have been found associated to production of haemolysin BL including, hemolysis, vascular permeability and necrosis (Beecher, 2000). A relation between differentiation of swarm cells, which require the activity of *fliY*, and the production of the L2 component of the haemolysin BL, was observed. *fliY* encode a motor-switch protein (FliY) important in the assembly of the flagellar filament (Senesi, 2002). After biofilm formation the majority of bacteria decrease the expression of flagellin (*fla* genes) (Houry, 2010). Several findings have led to hypothesize a connection between biofilm formation and virulence in *B. cereus*. Autoinducer 2 (AI-2) is a signaling molecule responsible for intra- and inter-species communication in response to cell density (Auger, 2006). When the autoinducer is added exogenously, the biofilm formation is inhibited as well as the dispersion of cells from preformed biofilm (Auger, 2006). Moreover, the transcriptional regulators, encoded by *plcR* and *codY*, also influences *B. cereus* biofilm formation and may be involved in the pathogenicity of this bacterium.

In fact, PlcR is the main regulator of extracellular virulence factors in *B. cereus,* and CodY is a repressor of biosynthesis of branched amino-acids involved in the attachment of cells (Auger, 2006; Fagerlund, 2014). Between the wide range of genes encoding virulence factors that are controlled by the pleiotropic regulator PlcR, the *nhe, hbl* and *cytK,* are particularly significant. However, the deletion of *plcR* does not eradicate the virulence of *B. cereus* which suggests that other factors are involved in bacterium pathogenicity (Tran, 2010). The presence of flagella has also been proved to be involved in virulence-related properties as well as, other PlcR-independent factor genes are involved in *B. cereus* pathogenicity, such as *inhA1, hlyII* and *ilsA* (Gohar, 2002; Tran, 2010). Biofilms produced by *B. cereus* are often associated with nutrient depletion and require the presence of biosurfactants (Hsueh, 2006; reviwed in Liaqat, 2013).

The synthesis of biosurfactants is also direct or indirectly regulated by the PlcR regulator. Biosurfactants reduce the surface tension and facilitated flagellum-independent sliding motility. Sporulation is also associated with *B. cereus* pathogenicity and biofilm formation, as referred above, in this case, the master transcription regulator Spo0A activates and triggers asymmetric sporulation division and transcription of the *spoIIA, spoIIE* and *spoIIG* loci that encode for development regulators. In addition SpoA and sigma-H (σ^H) are key regulators of the early steps of sporulation and biofilm formation. Strains of *B. cereus* with depletion of the transcription regulator Spo0A and σ^H are unable to form biofilm (reviwed in Liaqat, 2013).

Acknowledgments

AB holds an individual FCT fellowship (SFRH/BPD/98684/2013). JM holds a fellowship from the research project PTDC/DTP-SAP/1078/2012.

References

Andersson A, Granum PE et al. (1998) "The adhesion of *Bacillus cereus* spores to epithelial cells might be an additional virulence mechanism" in **International Journal of Food Microbiology,** 39:93-99

Auger S, Krin E et al. (2006) "Autoinducer 2 affects biofilm formation by *Bacillus cereus*" in **Applied and Environmental Microbiology,** 72:937-941

Auger S, Ramarao N et al. (2009) "Biofilm formation and cell surface properties among pathogenic and nonpathogenic strains of the *Bacillus cereus* group" in **Applied and Environmental Microbiology,** 75:6616-6618

Beecher DJ, Olsen TW et al. (2000) "Evidence for contribution of tripartite hemolysin BL, phosphatidylcholine-preferring phospholipase C, and collagenase to virulence of *Bacillus cereus* endophthalmitis" in **Infection and Immunity,** 68:5269-5276

Behling R, Eifert J et al. (2010) "Selected pathogens of concern to industrial food processors: infectious, toxigenic, toxico-infectious, selected emerging pathogenic bacteria" in **Principles of microbiological troubleshooting in the industrial food processing environment**. Kornacki JL, Springer New York:5-61

Bottone EJ (2010) "*Bacillus cereus,* a volatile human pathogen" in **Clinical Microbiology Reviews,** 23:382-398

Choi N-Y, Kim B-R et al. (2013) "Biofilm formation, attachment, and cell hydrophobicity of foodborne pathogens under varied environmental conditions" in **Journal of the Korean Society for Applied Biological Chemistry,** 56:207-220

Drobniewski FA (1993) "*Bacillus cereus* and related species" in **Clinical Microbiology Reviews,** 6:324-338

Fagerlund A, Dubois T et al. (2014) "SinR controls enterotoxin expression in *Bacillus thuringiensis* biofilms" in **PLoS ONE,** 9(4): e96707

Funada H, Uotani C et al. (1988) "*Bacillus cereus* bacteremia in an adult with acute leukemia" in **Japanese Journal of Clinical Oncology,** 18:69-74

Gohar M, Økstad OA et al. (2002) "Two-dimensional electrophoresis analysis of the extracellular proteome of *Bacillus cereus* reveals the importance of the PlcR regulon" in **Proteomics,** 2:784-791

Granum PE, Lund T (1997) "*Bacillus cereus* and its food poisoning toxins" in **FEMS Microbiology Letters,** 157:223-228

Houry A, Briandet R et al. (2010) "Involvement of motility and flagella in *Bacillus cereus* biofilm formation" in **Microbiology,** 156:1009-1018

Hsueh Y-H, Somers E et al. (2008) "Characterization of the codY gene and its influence on biofilm formation in *Bacillus cereus*" in **Archives of Microbiology,** 189:557-568

Hsueh YH, Somers EB et al. (2006) "Biofilm formation by *Bacillus cereus* is influenced by PlcR, a pleiotropic regulator" in **Applied and Environmental Microbiology,** 72:5089-5092

Jensen GB, Hansen BM et al. (2003) "The hidden lifestyles of *Bacillus cereus* and relatives" in **Environmental Microbiology,** 5:631-640

Kamar R, Gohar M et al. (2013) "Pathogenic potential of *Bacillus cereus* strains as revealed by phenotypic analysis" in **Journal of Clinical Microbiology,** 51:320-323

Kotiranta A, Lounatmaa K et al. (2000) "Epidemiology and pathogenesis of *Bacillus cereus* infections" in **Microbes and Infection,** 2:189-198

Kuroki R, Kawakami K et al. (2009) "Nosocomial bacteremia caused by biofilm-forming *Bacillus cereus* and *Bacillus thuringiensis*" in **Internal Medicine,** 48:791-796

Lee M-J, Ha J-H et al. (2010) "Reduction of *Bacillus cereus* contamination in biofilms on stainless steel surfaces by application of sanitizers and commercial detergent" in **Journal of the Korean Society for Applied Biological Chemistry,** 53:89-93

Liaqat I, Ahmed SI et al. (2013) "Biofilm formation and sporulation in *Bacillus subtilis*" in **International Journal of Microbiology Research and Reviews,** 1:61-67

McCabe-Sellers BJ, Beattie SE (2004) "Food safety: Emerging trends in foodborne illness surveillance and prevention" in **Journal of the American Dietetic Association,** 104:1708-1717

McLinden T, Sargeant J et al. (2014) "Component costs of foodborne illness: a scoping review" in **BMC Public Health,** 14:1-12

Nam H, Seo HS et al. (2014) "Efficacy of gaseous chlorine dioxide in inactivating *Bacillus cereus* spores attached to and in a biofilm on stainless steel" in **International Journal of Food Microbiology,** 188:122-127

Newell DG, Koopmans M et al. (2010) "Food-borne diseases — The challenges of 20 years ago still persist while new ones continue to emerge" in **International Journal of Food Microbiology,** 139, Supplement:S3-S15

Økstad OA, Kolstø AB (2011) "Genomics of *Bacillus* Species" in **Genomics of Foodborne Bacterial Pathogens**. Wiedmann M, Zhang W, Springer New York:29-53

Oosthuizen MC, Steyn B et al. (2002) "Proteomic analysis reveals differential protein expression by *Bacillus cereus* during biofilm formation" in **Applied and Environmental Microbiology,** 68:2770-2780

Pirttijärvi TSM, Andersson MA et al. (2000) "Properties of *Bacillus cereus* and other bacilli contaminating biomaterial-based industrial processes" in **International Journal of Food Microbiology,** 60:231-239

Ramarao N, Sanchis V (2013) "The pore-forming haemolysins of *Bacillus cereus*: a review" in **Toxins,** 5:1119-1139

Rasko DA, Altherr MR et al. (2005) "Genomics of the *Bacillus cereus* group of organisms" in **FEMS Microbiology Reviews,** 29:303-329

Ryu JH, Beuchat LR (2005) "Biofilm formation and sporulation by *Bacillus cereus* on a stainless steel surface and subsequent resistance of vegetative cells and spores to chlorine, chlorine dioxide, and a peroxyacetic acid–based sanitizer" in **Journal of Food Protection,** 68:2614-2622

Sauer K (2003) "The genomics and proteomics of biofilm formation" in **Genome Biology,** 4:219

Schoeni JL, Lee Wong AC (2005) "*Bacillus cereus* food poisoning and its toxins" in **Journal of Food Protection,** 68:636-648

Senesi S, Celandroni F et al. (2002) "Swarming motility in *Bacillus cereus* and characterization of a *fliY* mutant impaired in swarm cell differentiation" in **Microbiology,** 148:1785-1794

Senesi S, Salvetti S et al. (2010) "Features of *Bacillus cereus* swarm cells" in **Research in Microbiology,** 161:743-749

Simões M, Simoes LC et al. (2008) "Antagonism between *Bacillus cereus* and *Pseudomonas fluorescens* in planktonic systems and in biofilms" in **Biofouling,** 24:339-349

Stenfors ALP, Fagerlund A et al. (2008) "From soil to gut: *Bacillus cereus* and its food poisoning toxins" in **FEMS Microbiology Reviews,** 32:579-606

Tran S-L, Guillemet E et al. (2010) "CwpFM (EntFM) is a *Bacillus cereus* potential cell wall peptidase implicated in adhesion, biofilm formation, and virulence" in **Journal of Bacteriology,** 192:2638-2642

Wijman JGE, De Leeuw PPLA et al. (2007) "Air-liquid interface biofilms of *Bacillus cereus*: formation, sporulation, and dispersion" in **Applied and Environmental Microbiology,** 73:1481-1488

Xu YB, Chen M et al. (2014) "The phosphotransferase system gene ptsI in the endophytic bacterium *Bacillus cereus* is required for biofilm formation, colonization, and biocontrol against wheat sharp eyespot" in **FEMS Microbiology Letters,** 354:142-152

Chapter 13

Insights on Enterococcus biofilm factors from expression studies

Maria de Fátima Silva Lopes*
IBET- IBET, Apartado 12, 2781-901 Oeiras, Portugal
ITQB- Instituto de Tecnologia Química e Biológica, Universidade Nova de Lisboa, Av. Da República, 2780-157 Oeiras, Portugal
flopes@itqb.unl.pt

Contents of chapter

Introduction to the chapter

Opportunistic pathogens constitute true challenges for researchers and for those involved in health care practices. Their pathogenicity mechanisms are not just dependent on specific virulence factors, but host factors and environments also contribute to infectivity of opportunistic bacteria. Bacteria belonging to the *Enterococcus* genus are such an example. The genus *Enterococcus* comprises over 55 species and this number increases every year. They are natural colonizers of the human genitourinary tract since birth. However, they are also ale to cause infections, described for the first time in the 19th century (MacCallum, 1899). Their natural resistance to several antibiotics used in the clinical setting granted these bacteria with selective advantage and allowed them to become one of the prime causes of nosocomial infections worldwide (reviewed in Arias 2012). The fact that they are human commensals and also human pathogens able to resist current antimicrobial therapies urges us to find new approaches to fight and control enterococcal infections.

Despite recognized as a genus since 1984, it was only in the last 10 years that we have seen a boom in genetic and functional research on *Enterococcus*. The first enterococcal genome was sequenced and annotated in 2003 (Paulsen, 2003) and since then, several expression studies have attempted to identify genes responsible for the pathogenic potential of these bacteria. Enterococci are very harsh bacteria, able to stand extreme conditions of temperature, dryness, pH and salinity (reviewed in Ogier, 2008). They are able to survive in abiotic surfaces for long periods of time and are found in sand, soil and water. They are also used in fermented food products, namely cheese, where they contribute to flavour and control of pathogenic species through bacteriocin production (reviewed in Ogier, 2008; Arias, 2012). Inside the human host, they are able to survive the low pH in the stomach and colonize the gut. They can cross the epithelial barrier of the intestine (Wells, 1990), survive and multiply in different host tissues, where they stand chemical and cellular host defenses (Krueger, 2004). Therefore, many of the trasncriptomic studies have searched for genetic determinants involved in enterococcal survival to salt, bile, serum, urine, hydrogen peroxide, antibiotics and divalent metal ions. Although the overall conclusion is that virulence of these bacteria is multifactorial and dependent on the strain and host conditions, these studies have provided valuable clues on how these bacteria turn from commensal to pathogenic. In particular, and by mimicking host related environments, microarray and RIVET (recombinase-based *in vivo* expression technology) studies have identified enterococcal factors involved in biofilm formation. Ability to form biofilms is crucial for bacteraemia and urinary tract infections (UTI), two of the most severe and common enterococcal infections, respectively. Combining whole genome expression studies with mutant analysis have proven to be valuable tools in providing a solid knowledge base for developing new approaches to fight enterococcal infections.

Impact in clinical settings

Bacteria from the genus *Enterococcus* are capable of surviving in most of the environments surrounding humans, but are known to be part of the core microbiota in the human gut. In recent years, along with the increased interest in the gut microbiota, due to its contribution to many human diseases and disorders, we became aware of the importance and role of *Enterococcus* in human health as a result of its presence in such a multispecies community. On the one hand, its presence controls gut colonization by pathogens, but on the other hand, it constitutes an important route of entry into the blood and access to inner host tissues. These Gram-positive, homofermentative and facultative anaerobic bacteria are now considered as "serious" in the threat level scale of the USA CDC latest report on antimicrobial resistance surveillance (CDC, 2013). Despite known for more than a century as a causative agent of endocarditis (MacCallum, 1899), the success of enterococci as pathogens increased alongside with the use of antibiotics. This is due to their intrinsic resistance to many antibiotics (such as penicillin, clindamycin, aminoglicosides) (reviewed in Arias, 2012), to their ability to accumulate acquired antibiotic resistances (such as erythromycin, vancomycin, tetracycline, rifampicin, quinolones high level aminoglicosides) (reviewed in Arias, 2012), to their high adaptability to different environmental conditions and to their ability to survive for prolonged periods of time in medical devices and surfaces (reviewed in Arias, 2012). However, their ability to cause disease depends also on the health status of the host.. In fact, enterococci only cause disease in hospitalized human patients who have gone through surgical procedures, who have impaired immune systems or who have gone through prior antimicrobial therapy (reviewed in Arias, 2012). Enterococci rank now as the second most common organisms recovered from catheter-associated infections of the bloodstream and urinary tract, and from skin and soft-tissue infections (Arias, 2012). The species most frequently associated to nosocomial infections is *Enterococcus faecalis*. However, in recent years *Enterococcus faecium* resistant to vancomycin became almost as common as *E. faecalis* as a cause of nosocomial infections (Arias, 2012). The fact that enterococci are now considered a major clinical problem is related to the appearance of hospital adapted clones carrying multiple antibiotic resistance determinants (reviewed in Arias, 2012). Ten years have passed since the first *E. faecalis* genome was sequenced (Paulsen, 2003). Now, around 20 *E. faecalis* and 75 *E. faecium* genomes have been sequenced and include commensal and clinical strains, of both human and animal origin, with isolation dates ranging several decades (Palmer, 2010; Qin, 2012; Howden, 2013; Lebreton, 2013).

We are beginning to understand the reasons for the exponential increased relevance as nosocomial pathogens seen in the last forty years. *E. faecalis* and *E. faecium* species evolved differently. In the case of *E. faecalis*, there is a conserved core genome and the difference between clinical and commensal strains appears to be related to the accumulation of exogenous DNA such as plasmids, phages and transposons (Palmer, 2012). On the other hand, *E. faecium* adaptation to the hospital environment has come through accumulated mutations in the species core genome which provided advantages during antimicrobial therapy (Lebreton, 2013).

Biofilm associated enterococcal factors

Although antibiotic resistance has been the driving force for enterococcal evolution and research, recent availability of high throughput genetic tools, such as genome sequencing and transcriptomic analysis, has helped researchers to find bacterial factors relevant for infection (see table 13.1). Unlike streptococci and staphylococci, most enterococci do not produce a set of potent pro-inflammatory toxins. However, they are equipped with many genes encoding proteins that help them in adhesion and degradation of host proteins and tissues and in dealing with host immune defenses (reviewed in Jett, 1994; Sava, 2010; Arias, 2012).

Many of the identified *E. faecalis* virulence factors are also involved in biofilm formation and include surface proteins, proteases, lysins, enzymes with a role in metabolism, regulatory proteins, heat shock proteins and others (reviewed in Paganelli, 2012). The ability to form biofilms is a clinically relevant property of *E. faecalis* and contributes to increased tolerance to antibiotics and infection in the nosocomial environment. In fact, biofilms play a critical role in two of the most frequent and fatal enterococcal infections, UTI and endocarditis, respectively (Paganelli, 2012). Therefore, an effort has been put on finding enterococcal factors relevant to biofilm formation and on understanding which environmental conditions trigger expression of biofilm associated genes. Different techniques have been used and include mutant analysis (targeted and random) (Hancock, 2004; Kristich, 2008), qRT-PCR (Shepard, 2002; Hew, 2007; Carlos, 2010; Lenz, 2010) and RIVET (Ballering, 2009). The results from the two high throughput screenings targeting biofilm enterococcal factors (Kristich, 2008; Ballering, 2009), despite using the same enterococcal isolate (*E. faecalis* OG1RF), did not overlap, but allowed for the identification of 10 and 68, respectively, new biofilm–associated factors. Although over 100 biofilm associated enterococcal factors have been identified, not all were tested for a role in disease, colonization or infection and for those which were tested not all have proven to contribute to enterococcal pathogenicity. On top of that, different strains or animal models of disease also provide different roles for biofilm-associated factors, likely reflecting differences in infection development correlated with distinct roles in the different stages of biofilm development.

One of the best studied biofilm-associated *E. faecalis* factors are two proteases, gelatinase (GelE) and serine protease (SprE), encoded in one operon and regulated by a quorum-sensing activated regulatory system known as Fsr. The two proteases and their regulator have been implicated in virulence in several animal models of infection, in degradation of immune system antimicrobial peptides, translocation across epithelial barrier, degradation of human proteins (reviewed in Teixeira, 2013)) and in colitis in mice (Steck, 2011). GelE was first discovered as a proteolytic activity in 1964 (Bleiweis, 1964) and associated to caries in 1975 (Gold, 1975) but its gene was only identified 15 years later (Su, 1991). Fsr role as its regulator was later described (Qin, 2000; Nakayama, 2006) and is since then known to comprise four genes: *fsrA*, *fsrB*, *fsrC* and *fsrD*. The last encodes an auto-inducing cyclic peptide named gelatinase biosynthesis-activating pheromone (GBAP), and this peptide is processed and exported out of the cell by FsrB. Accumulation of GBAP outside cells is sensed by the FsrC histidine kinase, leading to the activation of the response regulator FsrA. Activated FsrA induces expression of the *fsrBDC* genes forming an auto regulatory circuit that results in a rapid, exponential increase in GBAP signaling and in expression of *gelE-sprE* operon. Virulence studies suggested that Fsr could be regulating other virulence factors in the *E. faecalis* genome.

This was confirmed by two recent transcriptomic studies, which addressed the role of Fsr and the two proteases in the overall gene expression in *E. faecalis* (Bourgogne, 2006; Teixeira, 2013). Fsr was found to regulate other genes, namely *ef1097*, putatively encoding a bacteriocin (Swe, 2007), in two genetically distinct *E. faecalis* strains, OG1RF (Bourgogne, 2006) and V583 (Teixeira, 2013). EF1097 was further studied and found to contribute to *E. faecalis* infection in *Drosophila melanogaster* animal model (Teixeira, 2013). These studies illustrate the impact of transcriptomic studies in unravelling virulence mechanisms in pathogens.

E. faecalis transcriptome studies

E. faecalis first genome was published in 2003 (Paulsen, 2003) and since then this species has been the focus of several transcriptomic studies, using microarray technology (Table 13.1). *E. faecalis* is known for its ability to stand harsh conditions, namely those experienced inside the host digestive tract. Therefore, transcriptomic studies addressed different environmental conditions, such as high salt concentrations, presence of bile salts or SDS. *E. faecalis* are able to cross the epithelial barrier of the intestine it colonizes and survive inside macrophage defense cells, allowing it to travel to distant host sites where it can initiate an infection (Wells, 1990; Krueger, 2004). Therefore, other transcriptomic studies addressed the ability of this pathogen to survive oxidative stress, high divalent metal ion concentrations, blood and urine. Its known intrinsic resistance to several antibiotics used in the clinical practice also led researchers to identify the overall *E. faecalis* response to antibiotics, namely erythromycin, vancomycin and cloramphenicol. More recently, with the advances in high throughput sequencing technologies, genes differentially expressed during *in vivo* colonization of mice were searched using

Table 13.1 | Transcriptomic studies and differential expression of enterococcal factors implicated in biofilm formation.

Experiment	Species/ strain	Biofilm-associated factors[c]	Differential expression[a]	Reference
<u>Environmental cue:</u>				
Erythromycin (50µg/ml)	*E. faecalis* V583	-	-	Aakra, 2005
Vancomycin (10µg/ml)	*E. faecalis* V583	*fsrABDC* *gelEsprE* *salB* *epaB* *dltABCD*	D D D U U	Ribeiro, 2014
Cloramphenicol (2.5µg/ml)	*E. faecalis* V583	-	-	Aakra, 2010
Urine	*E. faecalis* OG1RF	*bopABC/malP* *BM* *salA* *salB* *srtA* *atlA* *epa*	U D D U D D	Vebo, 2010
	E. faecalis MMH594	*salA* *salB* *srtA* *atlA* *epa*	D D U D D	
	E. faecalis Symbioflor1	*ef3314* *salA* *salB* *atlA* *epa*	U D D D D	
Blood[b] (10% and 100%)	*E. faecalis* V583	*dltAB* *epa*	D D	Vebo, 2009

		bopABC/malP *BM* *salB* *fsrABDC* *gelEsprE*	U D D D	
Bile[c] (1%)	*E. faecalis* V583	-	-	Solheim, 2007
SDS (0.06%)	*E. faecalis* V583	-	-	Solheim, 2007
NaCl (6.5%)	*E. faecalis* V583	*gelEsprE* *dnaK* *epa*	D U U	Solheim, 2014
Hydrogen peroxide (1.75mM)	*E. faecalis* V583 Ery^S	*ef3314*	U	Yan, 2014
Iron (0.5mM)	*E. faecalis* OG1RF	-	-	López. 2012
Copper (0.5mM)	*E. faecalis* OG1RF	-	-	Reyes-Jara, 2010
Copper (0.05mM)	*E. faecalis* V583	*dltABCD* *ef3314*	U U	Abrantes, 2011
Zinc (4mM)	*E. faecalis* V583	-	-	Abrantes, 2011
Manganese (0.4mM)	*E. faecalis* V583	-	-	Abrantes, 2011
<u>*In vivo*</u>:				
Intestine colonization in mice (one week)[d]	*E. faecalis* OG1RF	*gelEsprE*	D	Lindenstrauβ, 2014

[a], D, down-regulated; U, up-regulated. [b], horse blood. [c], bovine bile. [d], Using SOLiD sequencing analysis. [e], *salA*, secretory antigen A; *salB*, secretory antigen B; *srtA*, sortase; *epa* operon, enterococcal polysaccharide antigen; *dlt* operon, D-alanylation of LTA; *atlA*, autolysin A; *bop/mal* operon, biofilm on plastic surfaces/maltose metabolism; dnaK, heat shock protein K; EF3314, putative surface-exposed antigen; *gelE*, gelatinase; *sprE*, serine protease; *fsr* operon, *E. faecalis* sensor regulator.

SOLiD sequencing. Table 13.1 summarizes transcriptomic studies on *E. faecalis* (both microarray and sequence based) and identifies genes associated to biofilm formation which were differentially expressed under the studied conditions.

Interestingly, the *gelE*, *sprE* and *fsr* virulence factors were found to be down regulated in those expression studies where these genes were differentially expressed (Vebo, 2009; Ribeiro, 2014; Solheim, 2014; Lindenstrauβ, 2014). In particular, in the *in vivo* study, down-regulation of the protease genes was concordant with the absence of tissue injury (Lindenstrauβ, 2014). However, the number of bacterial counts was high enough to trigger the Fsr system, as it is known to occur in planktonic growth. The results from transcriptomic studies on *gelE-sprE* and other biofilm/infection associated factors, evidence the need for a better experimental design in transcriptomic study, better mimicking an in vivo infection situation.

One of the most relevant and puzzling aspects, surrounding enterococcal potential for virulence, is the fact that virulence factors are transversal to the species and are thus present in enterococci isolated from infections, form the human gut, from food and from abiotic environments. Moreover, these bacteria are necessary for a healthy human being, but are also able to cause infection and death. Therefore, the main question behind many research studies on enterococci is "what makes an isolate become virulent?" In order to find clues to solve this puzzle, several studies rely on comparison between commensal, food and clinical isolates, at the gene and expression levels. One of the transcriptomic studies mentioned in Table 13.1 compared three different *E. faecalis* strains: a laboratory strain derived from a human commensal isolate (OG1RF), a clinical isolate from a bloodstream infection (V583) and a probiotic strain (Symbioflor 1). Despite the genome differences between these three strains, they all revealed a similar ability to grow in human urine and an overall similar transcriptomic response, particularly concerning biofilm-associated factors. In general, results from this comparison suggest that the pathogenic potential of a certain *E. faecalis* strain may to a great extent be determined by the presence of fitness and virulence factors, rather than the level of expression of such traits (Vebo, 2010).

Overall, the presented transcriptomic studies have characterized the expression signature to different environmental conditions. Follow up studies, further exploring the results from microarray data, succeeded in identifying new virulence factors and genes involved in biofilm formation. Analysis of the transcriptomic response of *E. faecalis* to manganese, zinc and copper (Abrantes, 2011) allowed the identification of a motif in the promoter regions of genes repressed by manganese and induced by copper and zinc (Abrantes, 2013). EfaR regulator was found to be associated to the identified motif; and to play a role in several processes involved in host colonization and infection, namely biofilm formation (Abrantes, 2013). Also, further analysis of the transcriptomic responses to metal ions (Abrantes, 2011) identified a DNA binding motif in the promoter regions of genes strongly responding to zinc, and led to the identification of two new genes involved in the *E. faecalis* ability to stand human host defenses, such as high zinc concentrations, oxidative stress and lysozyme (Abrantes, 2014).

In summary, gene expression studies, though providing new virulence factors and valuable clues on pathogenic strategies of *E. faecalis,* also highlighted the need to properly mimic host conditions in order to understand particular expression patterns of virulence factors. The recent advances in high throughput sequencing techniques will certainly allow performing *in vivo* studies during infection and colonization in adequate animal host models. We can expect major breakthroughs in the next coming years.

References

Aakra A, Vebo H et al. (2005) "Transcriptional response of *Enterococcus faecalis* V583 to erythromycin" in **Antimicrobial Agents and Chemotherapy**, 49:2246-2259.

Aakra A, Vebo H et al. (2010) "The response of *Enterococcus faecalis* V583 to chloramphenicol treatment" in **International Journal of Microbiology**, 2010: (483048).

Abrantes MC, Lopes MF et al. (2011) "Imapct of manganese, copper and zinc ions on the transcriptome of the nosocomial pathogen *Enterococcus faecalis* V583" in **PlosOne**, 6(10): e26519.

Abrantes MC, Kok J et al. (2013) "EfaR is a major regulator of *Enterococcus faecalis* manganese transporters and influences processes involved in host colonization and infection" in **Infection and Immunity**, 81:935-944.

Abrantes MC, Kok J et al. (2014) "*Enterococcus faecalis* zinc-responsive proteins mediate bacterial defence against zinc overload, lysozyme and oxidative stress" in **Microbiology**, doi: doi: 10.1099/mic.0.080341-0

Arias CA, Murray BE (2012) "The rise of the *Enterococcus*: beyond vancomycin resistance" in **Nature Reviews Microbiology**, 10:266-278.

Ballering KS, Kristich CJ et al. (2009) "Functional genomics of *Enterococcus faecalis*: multiple novel genetic determinants for biofilm formation in the core genome" in **Journal of Bacteriology**, 191:2806-2814.

Bleiweis AS, Zimmerman LN (1964) "Properties of proteinase from *Streptococcus faecalis* var. *liquefaciens*" in **Journal of Bacteriology**, 88:653-659.

Bourgogne A, Hilsenbeck SG et al. (2006) "Comparison of OG1RF and an isogenic *fsrB* deletion mutant by transcriptional analysis: the Fsr system of *Enterococcus faecalis* is more than the activator of gelatinase and serine protease" in **Journal of Bacteriology**, 188:2875-2884.

Carlos AR, Semedo-Lemsaddeck T et al. (2010) "Transcriptional analysis of virulence-related genes in enterococci from distinct origins" in **Journal of Applied Microbiology**, 108:1563-1575.

CDC (2013) "Antibiotic resistance threats in the United States, 2103" in **Centers for Disease Control and Prevention,** USA.

Gold OG, Jordan HV et al. (1975) "The prevalence of enterococci in the human mouth and their pathogenicity in animal models" in **Archives in Oral Biology**, 20:473-477.

Hancock LE, Perego M (2004) "Systematic inactivation and phenotypic characterization of two-component signal transduction systems of *Enterococcus faecalis* V583" in **Journal of Bacteriology**, 186:7951-7958.

Hew CM, Korakli M et al. (2007) "Expression of virulence-related genes by *Enterococcus faecalis* in response to different environments" in **Systematic and Applied Microbiology**, 30:257-267.

Howden BP, Holt KE et al. (2013) "Genomic insights to control the emergence of vancomycin-resistant enterococci" in **mBio**, 4(4):e00412.

Jett BD, Huycke MM et al. (1994) "Virulence of enterococci" in **Clinical Microbiology Reviews**, 7:462-478.

Kristich CJ, Nguyen VT et al. (2008) "Development and use of an efficient system for random mariner transposon mutagenesis to identify novel genetic determinants of biofilm formation in the core *Enterococcus faecalis* genome" in **Applied and Environmental Microbiology**, 74:3377-3386.

Krueger WA, Krueger-Rameck S et al. (2004) "Assessment of the role of antibiotics and enterococcal virulence factors in a mouse model of extraintestinal translocation" in **Critical Care Medicine**, 32:467-471.

Lebreton F, van Schaik W et al. (2013) "Emergence of epidemic multidrug-resistant *Enterococcus faecium* from animal and commensal strains" in **mBio**, 4(4):e00534-13.

Lenz CA, Ferstl CMH et al. (2010) "Sub-lethal stress effects on virulence gene expression in *Enterococcus faecalis*" in **Food Microbiology**, 27:317-326.

Lindenstrauβ AG, Ehrmann MA et al (2014) "Transcriptome analysis of *Enterococcus faecalis* towards its adaptation to surviving in the mouse intestinal tract" in **Archives of Microbiology**, 196:423-433.

López G, Latorre M et al. (2012) "Transcriptomic response of *Enterococcus faecalis* to iron excess" in **Biometals**, 25:737-747.

MacCallum WG, Hastings TW (1899) "A case of acute endocarditis caused by *Micrococcus zymogenes* (nov.spec.), with a description of the microorganism" in **Journal of Experimental Medicine**, 4:521-534.

Nakayama J, Chen S et al. (2006) "Revised model for *Enterococcus faecalis fsr* quorum-sensing system: the small open reading frame *fsrD* encodes the gelatinase biosynthesis-activating pheromone propeptide corresponding to staphylococcal AgrD" in **Journal of Bacteriology**, 188:8321-8326.

Ogier J-C, Serror P (2008) "Safety assessment of dairy microorganisms: the *Enterococcus* genus" in **International Journal of Food Microbiology**, 126:291-301.

Paganelli FL, Willems RJ et al. (2012) "Optimizing future treatment of enterococcal infections: attacking the biofilm?" in **Trends in Microbiology**, 20:40-49.

Palmer KL, Carniol K et al. (2010) "High-quality draft genome sequences of 28 *Enterococcus* sp. isolates. **Journal of Bacteriology**, 192:2469 –2470.

Palmer KL, Godfrey P et al. (2012) "Comparative genomics of enterococci: variation in *Enterococcus faecalis*, clade structure in *E. faecium*, and defining characteristics of *E. gallinarum* and *E. casseliflavus*", in **mBio**, 3(1):e00318-11.

Paulsen IT, Banerjei L et al. (2003) "Role of mobile DNA in the evolution of vancomycin-resistant *Enterococcus faecalis*" in **Science**, 299:2071-2074.

Qin X, Weinstock GM et al. (2000) "The effects of *Enterococcus faecalis fsr* genes on production of gelatinase and a serine protease and virulence" in **Infection and Immunity**, 68:2579-2586.

Qin X, Galloway-Peña JR et al. (2012) "Complete genome sequence of *Enterococcus faecium* strain TX16 and comparative genomic analysis of Enterococcus faecium genomes" in **BMC Microbiology**, 12:135-154.

Reyes-Jara A, Latorre M et al. (2010) "Genome-wide transcriptome analysis of the adaptative response of *Enterococcus faecalis* to copper exposure" in **Biometals**, 23:1105-1112.

Ribeiro T, Teixeira N et al. (2014) "Transcriptomic study reveals new pathways and genes involved in *Enterococcus faecalis* V583 response to a therapeutic dose of vancomycin" in **Archives of Clinical Microbiology**, 5:No 1:2.

Shepard BD, Gilmore MS (2002) Differential expression of virulence-related genes in *Enterococcus faecalis* in response to biological cues in serum and urine" in **Infection and Immunity**, 70:4344-4352.

Solheim M, Aakra A et al. (2007) "Transcriptional responses of *Enterococcus faecalis* V583 to bovine bile and sodium dodecyl sulfate" in **Applied and Environmental Microbiology**, 73:5767-5774.

Solheim M, La Rosa SL et al. (2014) "Transcriptomic and functional analysis of NaCl-induced stress in *Enterococcus faecalis*" in **PlosOne**, 9(4):e94571.

Steck N, Hoffmann M et al. (2011) "*Enterococcus faecalis* metalloprotease compromises epithelial barrier and contributes to intestinal inflammation" in **Gastroenterology**, 141:959-971.

Su YA, Sulavik MC et al. (1991) "Nucleotide sequence of the gelatinase gene (*gelE*) from *Enterococcus faecalis* subsp. *liquefaciens*" in **Infection and Immunity**, 59: 415-420.

Teixeira N, Varahan S et al. (2013) "*Drosophila* host model reveals new *Enterococcus faecalis* quorum-sensing associated virulence factors" in **PlosOne**, 8(5):e64740.

Vebo HC, Snipen L et al. (2009) "The transcriptome of the nosocomial pathogen *Enterococcus faecalis* V583 reveals adaptive responses to growth in blood" in **PlosOne**, 4(11):e7660.

Vebo HC, Solheim M et al. (2010) "Comparative genomic analysis of pathogenic and probiotic *Enterococcus faecalis* isolates, and their transcriptional responses to growth in human urine" in **PlosOne**, 5(8):e12489.

Wells CL, Jechorek RP et al. (1990) "Evidence for the translocation of *Enterococcus faecalis* across the mouse intestinal tract" in **Journal of Infectious Diseases**, 162: 82–90.

Yan X, Budin-Verneuil A et al. (2014) "Transcriptomic response of *Enterococcus faecalis* V583 to low hydrogen peroxide levels" in **Current Microbiology**, doi 10.1007/s00284-014-0691-8.

Chapter 14

Candida biofilms gene determinants

Mariana Henriques*, Maria Elisa Rodrigues, Sónia Silva
CEB, Centre of Biological Engineering, LIBRO – Laboratório de Investigação em Biofilmes Rosário Oliveira, Universidade do Minho, Braga, Portugal
* **mcrh@ceb.uminho.pt**

Contents of chapter

Introduction to the chapter

Infections caused by *Candida* species have increased dramatically in recent years. *Candida albicans* is considered to be the major pathogenic agent of candidiasis. However, other non-*Candida albicans Candida* (NCAC) species have been identified as potential human pathogens, namely, *Candida glabrata, Candida tropicalis* and *Candida parapsilosis.* The pathogenicity of *Candida* species is described to be facilitated by a large number of virulence factors with the most important being the ability to adhere to medical devices and/or host cells and the easiness to form biofilms. These *Candida* species biofilms, besides the high microbial charge associated, have a high resistance to antifungal agents, making treatment and eradication very difficult. Moreover, recent genomic DNA sequencing projects have suggested the involvement of important genes in *Candida* species biofilm formation and antifungal resistance and this will be disclosed in this chapter.

Candida and Candidiasis

In recent years, the number of infections caused by *Candida* species, called Candidiasis, has increased, being the fourth cause of health care-associated infections (reviewed in Silva, 2012; reviewed in Rodrigues, 2014). This rise in *Candida* infections is usually associated with the increased use of medical devices such as implants, dentures and catheters, the overuse of antimicrobial agents, the aging population and the increasing number of immunocompromised patients (reviewed in Kojic, 2004).

Candidiasis can be acquired by endogenous or exogenous pathways. The first one is usually when patients have *Candida* species in their microbial flora and by some host alteration (immunodeficiency, surgery, transplantation, diabetes or antimicrobial therapy) these yeasts proliferate, invade tissues and cause infections. On the other hand, exogenous pathways occurs by transport of *Candida* from a contaminated surface, such as a medical device, to a local not yet colonized, resulting in candidiasis (Samaranayake, 2002). Candidiasis may also be classified as superficial or systemic. In fact, *Candida* species can colonize the skin of healthy people causing superficial infections. The superficial candidiasis can occur in practically any skin surface and is expressed in similar form to eczema between toes, toenails and skin lesions (Samaranayake, 2002). Another example of such infections is diapers dermatitis in children, where yeast proliferates due to the hot and humid conditions in the diaper. Concerning systemic infections, these may impair several organs, causing candidemia (bloodstream infection), candiduria (urinary tract infection), pulmonary candidiasis, endocarditis, nephritis, phlebitis, arthritis, osteomilite, endophthalmitis and lesions of the central nervous system (Kauffman, 2000).

Candidiasis is usually attributed to *C. albicans*. However, currently other species of NCAC such as *C. glabrata*, *C. tropicalis* and *C. parapsilosis* have emerged. This is related not only to the new identification methods, which allow a better distinction of species, but also to the excessive use of antifungal agents observed today, which is responsible for the emergence of resistance (Chakrabarti, 2009).

As mentioned, *C. albicans* is considered the major species responsible for candidiasis, being the most well studied so far. This species is dimorphic, capable of being in the unicellular or filamentous form, as hyphae or pseudohyphae. *C. albicans* has been associated with all kinds of candidiasis, since it can be found in all parts of the body (reviewed in Silva, 2012).

In recent years, *C. glabrata* has been described as having rapid dissemination in the organisms and associated with high mortality. This species has been very resistant to most antifungals applied in clinical practice. *C. glabrata* infections are more significant in the northern hemisphere, mainly in adults and elderly people. These yeasts are often isolated in the oral cavity, and sometimes are found in association with *C. albicans* (reviewed in Silva, 2012; reviewed in Rodrigues, 2014).

Candida tropicalis has been classified as the second or third leading cause of candidemia, depending on the country, especially in immunocompromised patients admitted to intensive care units. This species is also often associated with high rates of candiduria (reviewed in Negri, 2012).

Epidemiological data also show an increase in *C. parapsilosis* incidence as a candidemia agent, especially in neonatal intensive care units, due to its easy development on a medium rich in glucose as parental nutrition. This species is also frequently related to the lack of measures of hygiene of healthcare professionals, namely with lack of compliance with the rules of hands washing and disinfection (Bonassoli, 2005). In addition, *Candida krusei*, *Candida dubliniensis*, *Candida polished*, *Candida guilliermondii* and *Candida kefir* could also be mentioned as being responsible for Candidiasis.

A common aspect among the several *Candida* species is that they all have more than one virulence factor. Among the many virulence factors, the dimorphism, production of extracellular enzymes (such as proteases, phospholipases and hemolysins), ability to adhere to abiotic (medical devices) and biotic (host cells) surfaces and the ability to form biofilms can be mentioned (reviewed in Silva, 2011).

Candida Biofilms

Biofilm formation is directly associated to the ability to cause infections, as they are responsible for increased resistance of *Candida* species to traditional antifungal agents and act as a protection against the host immune response. Furthermore, the increasing and frequent use of medical devices such as catheters, dental, hip and knee prostheses, and intrauterine devices which are ideal surfaces for colonization, is directly related to the increase in biofilm formation and consequent increase of candidiasis cases (reviewed in Donlan, 2002). Thus, *Candida* biofilms have a big impact both in the hospital and in the patient's health, since they are associated to very high hospital costs, either due to a reduction of the effective time of use of these medical materials or to the prolongation of hospitalization of the patient, and to infections that can lead to death in individuals with compromised immune systems (Tumbarello, 2012).

Biofilms are referred to as "infections reservoirs", since they are a source of further infection and the main cause of high morbidity and mortality hospital rates. In fact, biofilms formed by clinical isolates of *C. albicans*, *C. tropicalis*, *C. glabrata* and *C. parapsilosis* have been associated with a higher mortality rate compared with isolates of the same species unable to form biofilm (reviewed in Negri, 2012; reviewed in Silva, 2012; reviewed in Rodrigues, 2014).

Infections associated to biofilms

Infections caused by *Candida* biofilms are of utmost importance due to the difficulty of treatment, especially when biofilms are associated with other microorganisms. The number of reports of infections associated to biofilms of different *Candida* species is becoming increasingly common, mainly because these microorganisms are part of the human flora.

Vulvovaginal candidiasis, an infection caused by the overgrowth of *Candida* in the mucosa of female genital tract, increasingly prevalent in the last decades in healthy young women, has been related to the use of vaginal contraceptives such as intrauterine devices, due to the ability of *Candida* species to form extensive biofilms on these devices (Foxman, 2013). The main agent responsible for this type of infections is *C. albicans*, but recent studies have shown increased NCAC species in vulvovaginal candidiasis (Foxman, 2013).

The use of central venous and urinary catheters is often associated to the development of biofilms. Urinary catheters are easily colonized by *Candida* species that have the ability to grow and form biolfims in urine, which represents a particularly danger, since the yeast cells present in biofilm may be released and re-colonize the bladder causing candidurias (reviewed in Negri, 2012).

Endocarditis is related to extensive biofilm formation on endothelium leading to obstruction of the heart valves. In addition to *C. albicans* species, *C. parapsilosis* species are often associated with these infections in high mortality rates (Martin, 2003).

The infections associated to oral, hip and knee prostheses have also been attributed to the presence of *Candida* species biofilm. These infections are almost responsible for graft rejection and subsequent replacement of these prostheses. Furthermore, the mechanical properties of these materials are also modified when biofilm is formed on its surface. It has been shown that mixed biofilms of *C. albicans* and *Streptococcus mutans* reduced the friction surfaces of titanium, also changing its corrosion potential (reviewed in Williams, 2011).

Structure and composition of biofilms

Biofilms are described as communities of microorganisms embedded in an extracellular matrix produced by cells. As in bacterial biofilms, after the initial phase of *Candida* adhesion to surfaces, cell division and proliferation occur, which result in mature biofilm formation composed of either yeast, hyphae or pseudo-hyphae depending on the species (Figure 14.1).

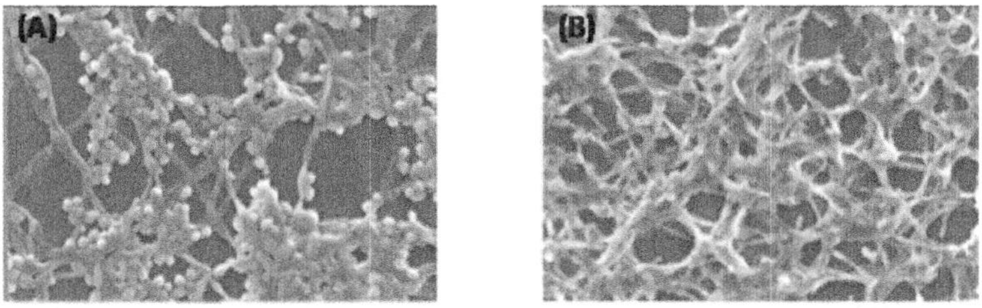

Figure 14.1 | *Candida* biofilm structure. Scanning electron microscopy images of *C. albicans* biofilm at (A) 24 h and (B) 48 h.

Interestingly, in the last years, reports demonstrate that biofilm formation by *Candida* species is dependent on the species, strain and environmental conditions (pH, medium composition, presence of oxygen) (reviewed in Silva, 2012). For example, in general, *C. glabrata* strains form lower amounts of biofilms in the presence of rich media (with high glucose levels), than *C. tropicalis* and *C. parapsilosis* (Silva, 2009). However, in artificial urine, these *C. glabrata* strains showed more ability to produce biofilms (nutrient-poor medium), than *C. tropicalis* and *C. parapsilosis* (Silva, 2010). These results confirm the inherent physiological differences between species and strains of *Candida* and alert to their different potential as pathogens.

C. albicans biofilms are generally composed by two distinct zones: a thin basal zone comprising yeasts and an upper thicker and less compact zone formed by filamentous forms (hyphae and/or pseudo-hyphae). In contrast with *C. albicans, C. parapsilosis* biofilms are thinner, less structured and formed mostly by yeast cells form. Clinical isolates of

C. tropicalis were recently classified as good biofilm producers and their biofilms consist of a cluster of cells and filamentous forms. Conversely, *C. glabrata* biofilms show a more compact morphology compared to *C. parapsilosis* and *C. tropicalis* biofilms (Silva, 2009).

As mentioned, biofilm cells are embedded in an extracellular matrix. The matrix of *C. albicans* biofilms is formed mainly of carbohydrates, proteins, phosphate and hexosamines (Baillie, 2000). Comparing the matrices of NCAC species biofilms, it is interesting to note that the *C. glabrata* extracellular biofilm matrix consists of high concentration of protein and carbohydrates and *C. parapsilosis* matrix by a high concentration of carbohydrates and low residual proteins (Silva, 2009). For *C. tropicalis*, the biofilm extracellular matrix has notable differences compared to other *Candida* species, containing a low content of carbohydrates and proteins. However, a major *C. tropicalis* matrix component is hexosamine (Al-Fattani, 2006). In addition, it was shown that *C. tropicalis* biofilms are more easily detached after lipase type VII and chitinase treatment and *C. albicans* are more easily detached after proteinase K and DNase(s) treatment, showing the differences in composition of the biofilm matrices (reviewed in Silva, 2012).

Genetic determination of Candida biofilm formation

As it has often been mentioned, the first step in biofilm formation is the adhesion of microorganisms to medical and host surfaces (reviewed in Silva, 2011). Similarly to the bacteria, yeasts of the *Candida* genus also adhere to the materials either by specific processes (binding proteins) or by nonspecific processes (hydrophobicity and surface charge) (reviewed in Silva, 2011). These proteins, called adhesins, are an integral part of the yeast cell wall. For *C. albicans* several genes are described coding adhesins, namely the families' manoproteins (MP), Agglutinin-Like Sequence (ALS) (reviewed in Hoyer, 2001). It is known that these genes are differentially regulated by relevant yeast physiological mechanisms, such as growth conditions. This difference in regulation and genetic variability found among *C. albicans* strains conditions the cell wall composition and consequently has an active role in the yeast pathogenicity (reviewed in Williams, 2011).

Adhesion of *C. albicans* to host cells is a key virulence mechanism and is associated with expression of Hyphal Wall Protein (*HWP)* gene. Alves (2014) using a qRT-PCR methodology showed that in co-infection of *C. albicans* and *C. glabrata, C. albicans* exhibited up-regulation of *HWP1* compared with single infection. Based on the qRT-PCR findings it is tempting to speculate that reconstituted human epithelium colonisation and invasion by *C. glabrata* cells is enhanced by the presence of *C. albicans* hyphae following *HWP1* up-regulation. In *C. glabrata* there are 23 adhesins described and the main group are encoded by the Epithelial Adhesion (*EPA*) genes family. It is also know that mutants in the *EPA1* gene have shown a reduced adhesion in vitro. In addition, the *EPA6* gene is not expressed by *C. glabrata in vitro,* but is produced *in vivo* during urinary infections, showing the adaptation of this species to environmental conditions to increase its adhesion (De Las Peñas, 2003). In *C. parapsilosis*, *ALS* and related protein 30 (Pga30) genes family were described as being associated to its adhesion (Butler, 2009). For *C. tropicalis*, at least three of the *ALS* gene family was identified (reviewed in Hoyer, 2001). Transcription factors play fundamental roles in both positive and negative regulation of biofilm formation through regulation of hyphal formation, cell surfaces properties responsible for adherence, matrix production and antifungal resistance (Dwivedi, 2011). *BCR1* (biofilm and cell wall regulator), a C_2H_2 zinc finger transcription factor, is a critical determinant of *C. albicans* biofilm formation in all environments studied to date. *BCR1* seems to be a conserved regulator of biofilm formation, because the *BCR1* ortholog of *C. parapsilosis* is required for biofilm formation as well (Ding, 2007). Recent work revealed that *C. parapsilosis BCR1* gene expression increased with the induction of biofilm formation by glucose. *ACE2*, another C_2H_2 zinc finger transcription factor, also contributes to *C. albicans* biofilm formation, probably through its role in adherence as well as hypha formation (Mulhern, 2006).

The *C. albicans* transcription factor *EFG1* (Enhanced Filamentous Growth), a global regulator of cell surface protein genes and hyphal formation, is required for biofilm formation as well (Finkel, 2011). The orthologs or best hits of *BCR1*, *ACE2*, and *EFG1*, including *C. glabrata* GACL0E06116g, GAGL0M04323g, and GAGL0L01771G, and *C. parapsilosis* CPAG00564, CPA00148, and CPAG00178, are stronger candidates for biofilm regulators in NCAC species (Fanning, 2012).

Additionally, many cell wall genes are induced in the biofilm environment. Altered expression of genes for ß-1,3 glucan synthesis and modification are features of *in vitro* and *in vivo* *C. albicans* biofilms including two gluconossyltransferases *BGL2, XOG1* and a glucanase *PHDR1*. *FSK1* (glucan synthase) has been shown also previously to exhibit and important role on *C. albicans* and *C. parapislosis* biofilm formation due to responsibility for the catalytic subunit of β-1,3 glucan synthase, which substrate is glucose (Nett, 2009). Other important zinc-response transcription factor, *ZAP1* is described also a negative regulator of matrix composition, soluble □-1,3 glucan, *in vivo* and *in vitro* *C. albicans* biofilms (Nobile, 2009). Given the connection between the ß-glucan production and matrix production, these genes may also play a role in matrix production.

Gene determinants in biofilm resistance

The classification of antifungal agents is based on their action mode. The polyenes group (e.g. amphotericin B) are fungicides due to their ability to interact with ergosterol causing pores in the cell membrane and consequently, causing loss of cellular contents and consequent cell death (Akins, 2005). The azoles group, the most widely used antifungal class at clinical level, act by inhibiting ergosterol synthesis by interfering with the enzyme lanosterol demitilase (Akins, 2005). 5-flucytosine is another group of antifungal agents that interacts with RNA molecules and consequently in protein synthesis. Echinocandins, a new class of antifungals, act at the level of the components of the cell wall of yeast by interfering with the synthesis of β-1,3-glucan (Pfaller, 2012).

In a clinical perspective, the main important characteristic of *Candida* biofilms is their role in increasing resistance to antifungal agents (Mukherjee, 2004). The reduced susceptibility of *Candida* biofilms to an antifungal (fluconazole) was first described in *C. albicans* in 1995. However, several research groups have demonstrated that there is an increased resistance to traditional antifungal agents (fluconazole and amphotericin B) of *Candida* cells in biofilm compared to the cells in planktonic growth (Fonseca, 2014). Biofilms of NCAC species such as *C. tropicalis*, *C. glabrata* and *C. parapsilosis* have also demonstrated antifungal resistance (reviewed in Rodrigues, 2014). It is known that the minimal inhibitory concentrations of the most common antifungal agents needed to reduce *Candida* biofilm biomass are about 10 times higher than the concentrations necessary to eliminate planktonic cells (reviewed in Donlan, 2002). Among the many mechanisms of tolerance to antifungal agents are the presence of extracellular matrix that acts as a diffusion barrier and as a neutralizing agent, and the presence of cells called "persisters" or less active cells, wherein the agent fails to act. The mechanisms underlying the resistance of biofilms to the antifungal agents' mechanisms are not clearly established yet. However, upregulation of mult-drug resistance transporter genes is common in *C. albicans* and *C. glabrata* (*MDR1*, *CDR1*, *CDR2*) biofilms *in vitro* (Nett, 2009). *Candida albicans* *MDR1* and *CDR2* are upregulated in *in vivo* biofilms, as *PDR16*, which is increased in fluconazole-resistant cells that over express *CDR1* and *CDR2*. Additionally, genes involved sterol biosynthesis is upregulated in *C. albicans* biofilms. Increases in *ERG* gene expression as well as multi-drug resistance transporters have been correlated with the increased azole resistance in *C. albicans* patient isolate samples, though their contribution to biofilm specific azole resistance (Fanning, 2012).

Conclusions

Candida species infections are frequently associated to biofilm formation, due to the increasing use of medical devices. Besides high cell density, these biofilms constitute a serious health problem due to antifungal agents resistance and due to its ability to evade the immune system response. These biofilms of different *Candida* species have a high diversity of matrix structure, composition and formation capability. This diversity is not only dependent on the species but also on the strain and environmental conditions. Thus, it is important to identify new *Candida* biofilm gene determinants to each species and draw new forms to combat biofilm formation and consequently the increasing number of candidiasis.

Acknowledgments

MER holds an individual FCT fellowship (SFRH/BPD/95401/2013).

References

Akins RA (2005) "An update on antifungal targets and mechanisms of resistance in *Candida albicans*" in **Medical Mycology**, 43:285-318

Al-Fattani MA, Douglas LJ (2006) "Biofilm matrix of *Candida albicans* and *Candida tropicalis*: chemical composition and role in drug resistance" in **Journal of Medical Microbiology**, 55:999-1008

Alves CT, Wei X-Q *et al.* (2014) "*Candida albicans* promotes invasion and colonisation of *Candida glabrata* in a reconstituted human vaginal epithelium" in **Journal of Infection**, 69:396-407

Baillie GS, Douglas LJ (2000) "Matrix polymers of *Candida* biofilms and their possible role in biofilm resistance to antifungal agents" in **Journal of Antimicrobial Chemotherapy**, 46:397-403

Bonassoli LA, Bertoli M *et al.* (2005) "High frequency of *Candida parapsilosis* on the hands of healthy hosts" in **Journal of Hospital Infection**, 59:159-162

Butler G, Rasmussen MD *et al.* (2009) "Evolution of pathogenicity and sexual reproduction in eight *Candida* genomes" in **Nature**, 459:657-662

Chakrabarti A, Chatterjee SS *et al.* (2009) "Recent experience with fungaemia: change in species distribution and azole resistance" in **Scandinavian Journal of Infectious Diseases**, 41:275-284

De Las Peñas A, Pan S-J *et al.* (2003) "Virulence-related surface glycoproteins in the yeast pathogen *Candida glabrata* are encoded in subtelomeric clusters and subject to RAP1- and SIR-dependent transcriptional silencing" in **Genes & Development**, 17:2245-2258

Ding C, Butler G (2007) "Development of a gene knockout system in *Candida parapsilosis* reveals a conserved role for BCR1 in biofilm formation" in **Eukaryotic Cell**, 6:1310-1319

Donlan RM, Costerton JW (2002) "Biofilms: survival mechanisms of clinically relevant microorganisms" in **Clinical Microbiology Reviews**, 15:167-193

Dwivedi P, Thompson A *et al.* (2011) "Role of Bcr1-activated genes Hwp1 and Hyr1 in *Candida Albicans* oral mucosal biofilms and neutrophil evasion" in **PLoS ONE** 6:e16218

Fanning, S Mitchell AP (2012) "Fungal biofilms" in **PLoS Pathogens**, 8:1-4

Finkel JS, Mitchell AP (2011) "Genetic control of *Candida albicans* biofilm development" in **Nature Reviews Microbiology**, 9:109-118

Fonseca E, Silva S *et al.* (2014) "Effects of fluconazole on *Candida glabrata* biofilms and its relationship with ABC transporter gene expression" in **Biofouling**, 30:447-457

Foxman B, Muraglia R *et al.* (2013) "Prevalence of recurrent vulvovaginal candidiasis in 5 European countries and the United States: results from an internet panel survey" in **Journal of Lower Genital Tract Disease**, 17:340-345

Hoyer LL (2001) "The ALS gene family of *Candida albicans*" in **Trends in Microbiology**, 9:176-180

Kauffman CA, Vazquez JA *et al.* (2000) "Prospective multicenter surveillance study of funguria in hospitalized patients" in **Clinical Infectious Diseases**, 30:14-18

Kojic EM, Darouiche RO (2004) "*Candida* infections of medical devices" in **Clinical Microbiology Reviews**, 17:255-267

Martin GS, Mannino DM *et al.* (2003) "The epidemiology of sepsis in the United States from 1979 through 2000" in **New England Journal of Medicine**, 348:1546-1554

Mukherjee PK, Chandra J (2004) "*Candida* biofilm resistance" in **Drug Resistance Updates**, 7:301-309

Mulhern SM, Logue ME *et al.* (2006) "*Candida albicans* transcription factor Ace2 regulates metabolism and is required for filamentation in hypoxic conditions" in **Eukaryotic Cell**, 5:2001-2013

Negri M, Silva S *et al.* (2012) "Insights into *Candida tropicalis* nosocomial infections and virulence factors" in **European Journal of Clinical Microbiology & Infectious Diseases**, 31:1399-1412

Nett JE, Lepak AJ *et al.* (2009) "Time course global gene expression analysis of an *in vivo Candida* biofilm" in **Journal of Infectious Diseases**, 200:307-313

Nobile CJ, Nett JE **et al.** (2009) "Biofilm Matrix Regulation by *Candida albicans* Zap1" in **PLoS Biology**, 7:e1000133

Pfaller MA, Castanheira M *et al.* (2012) "Frequency of decreased susceptibility and resistance to echinocandins among fluconazole-resistant bloodstream isolates of *Candida glabrata*" in **Journal of Clinical Microbiology**, 50:1199-1203

Rodrigues CF, Silva A *et al.* (2014) "*Candida glabrata*: a review of its features and resistance" in **European Journal of Clinical Microbiology & Infectious Diseases**, 33:673-688

Samaranayake LP, Fidel PL *et al.* (2002) "Fungal infections associated with HIV infection" in **Oral Diseases**, 8:151-160

Silva S, Henriques M *et al.* (2009) "Biofilms of non-*Candida albicans Candida* species: quantification, structure and matrix composition" in **Medical Mycology**, 47:681-689

Silva S, Negri M *et al.* (2010) "Silicone colonization by non-*Candida albicans Candida* species in the presence of urine" in **Journal of Medical Microbiology**, 59:747-754

Silva S, Negri M *et al.* (2011) "Adherence and biofilm formation of non-*Candida albicans Candida* species" in **Trends in Microbiology**, 19:241-247

Silva S, Negri M *et al.* (2012) "*Candida glabrata, Candida parapsilosis* and *Candida tropicalis*: biology, epidemiology, pathogenicity and antifungal resistance" in **FEMS Microbiology Reviews**, 36:288-305

Tumbarello M, Fiori B *et al.* (2012) "Risk factors and outcomes of candidemia caused by biofilm-forming isolates in a tertiary care hospital" in **PLoS ONE**, 7:e33705

Williams DW, Kuriyama T *et al.* (2011) "*Candida* biofilms and oral candidosis: treatment and prevention" in **Periodontology 2000**, 55:250-265

Glossary

Aggregation: the binding of genetically distinct microorganisms, in the planktonic phase, through specific interactions.

Adhesins: are cell-surface components that facilitate the adhesion to other cells or to surfaces.

Adhesion: the first step on biofilm formation.

Aggregation: the binding of genetically distinct microorganisms, in the planktonic phase, through specific interactions.

Amplicons: product of he DNA amplification by PCR.

Amplification: process of multiplying a DNA sequence.

Antifungal: is a drug used to treat and prevent Candida infections.

Antimicrobial compounds: substances that kill or inhibit the growth of microorganisms, such as bacteria, fungi or protozoans.

Antimicrobial resistance: tolerance of a microorganism to an antimicrobial drug that was originally effective for treatment of infections caused by it.

Bacitracin: an antibiotic.

Bacteraemia: presence of viable bactéria in the blood stream.

Bacterial sporulation: is the generation of a spore by a bacterium resulting in morphological conversion.

Bacteriophage/phage: virus that is able to infect bacteria.

Barcodes: specific sequences that are used to identify a specific sample.

Beta-galatosidase: enzyme that is encoded by the lacZ gene. It catalizes the hydrolysis of Lactose and other chromogenic substrates such as O-nitrophenyl-b-galactoside (ONPG) or Bromo-Chloro-Indolyl-Galactopyranoside (X-Gal) as an indicator of activity.

Biofilm physiology: structure, architecture and organization of thecells inside the biofilm.

Biofilm: microbial aggregates adhered to a surface (biotic or abiotic) and involved by protective extrapolymeric substances such as polysaccharides, proteins, lipids and extracellular DNA.

Biosurfactant: amphiphilic compound produced on living surfaces (mostly microbial cell surfaces) or secreted, which contains hydrophobic and hydrophilic moieties that reduce surface tension and interfacial tensions between individual molecules at the surface and interface, respectively.

Candidemia: Candida bloodstream infection.

Candidiasis: is an infection caused by any species of the genus Candida.

Candiduria: Candida urinary tract infection.

Chromosomes: organized structure of DNA and/or protein found in living cells.

Clinical isolates: microorganisms that were isolated directly from a patient suffering from a particular infection.

Codon: sequences of 3 nucleotides that encode for the synthesis of specific proteins, signal initiation or end of transcription.

Commensal microorganism: bacteria that is involved on a symbiotic relationship where the bacteria is benefited and the other organisms in neither benefited nor harmed.

Competence: the ability of cells to uptake exogenous DNA.

Complementary regions: two regions of DNA that are matched as they contain the mutually complementary sequences of nucleotides (for example for the region ATTCGA, the complementary region would have to be TAAGCT).

Cycle threshold: in qPCR, is the cycle number at which the fluorescence generated within a reaction reaches a fluorescent signal significantly above the background fluorescence. At the threshold cycle, a detectable amount of amplicon product has been generated. The threshold cycle is inversely proportional to the original relative expression level of the gene of interests, and therefore is a common unit of measurement of gene expression.

Cystic fibrosis: an inherited genetic disorder caused by a mutation in the gene which encodes the cystic fibrosis transmembranar conductance regulator protein (CFTR). This mutation leads to a greater predisposition to bacterial infection in patients with cystic fibrosis.

Cytoplasm: all the fluid inside the prokaryotic cell that surrounds all the functional molecules.

Cytotoxicity: quality or property of being toxic and even mortal to cells.

Denaturation (DNA): separation of the double strand into two single strands.

Determinant: a spatially localized molecule that causes cells to adopt a particular fate or set of related fates.

Diarrhoeal syndrome: is characterized by profuse diarrhea with abdominal cramps, but rarely vomiting or fever due to preformed enterotoxins, which begin 8 to 16 hours.

DNA: Deoxyribonucleic acid or DNA is the primary genetic information source and is present in all cells and some viruses.

DNA double-helix: consists of two right-handed polynucleotide chains that are coild about the same axis.

DNA polymerase: enzyme that helps catalyse the polymerization of nucleotides into a DNA strand.

DNA primase: catalyses the synthesis of a short RNA segment (called RNA primer) complementary to the ssDNA.

dNTP: or deoxyribonuclotide is the monomer, or single unit of DNA and RNA.

Dogma of molecular biology: is the order of information transfer of genetic information, DNA to RNA to proteins.

Downregulation: is the process by which the cells decrease the quantity of a cellular component.

Efflux pumps inhibitors: a class of compounds that act as inhibitors to the action of efflux pumps in cells. These pumps can actively transport antibiotics out of the cell, thereby providing resistance of the cell to the antibiotic.

Emetic syndrome: is characterized by nausea, vomiting, abdominal cramps and sometimes diarrhea caused by emetic toxin and has a short incubation period (1 to 5 hours).

Emetic toxin: is a highly stable peptidase of less than 10 KDa which is thermostable and resistant to proteolytic degradation. The emetic toxin causes a vomiting-type of food poisoning syndrome.

Endocarditis: is an inflammation of the inner layer of the heart, the endocardium.

Endophthalmitis: is an infection of the eyes caused by introduction of bacteria after trauma or surgery that result in severe inflammation and vision loss.

Etiological agent: a viable microorganism or its toxin that causes or may cause human disease or infection.

Eukaryotes: cells that contain complex structures within membranes, some examples are fungi, mammalian cells or plant cells.

Fastidious microorganism: microorganism that have unusual and/or complex nutritional needs and must be grown on enriched media.

Fimbria: a flagellum-like structure in bacteria that allows them to adhere to one another, to eukaryotic cells and even abiotic surfaces.
Foodborne illness: any illness resulting from the consumption of food.
Gene: is a segment of DNA involved in producing a polypeptide chain; it includes regions preceding and following the coding region (leader and trailer) as well as intervening sequences (introns) between individual coding segments (exons).
Genetic information: information encoded in genetic material within all living cells that serves as blueprint for each cell.
DNA sequencing: process of identification the order of the DNA nucleotides in a genome.
Genome: is the name given to the totality of the genetic material of a cell, it can be made up of DNA or RNA (in the case of some viruses).
Helicase: are motor proteins that move directionally along the nucleic acid phosphodiester backbone of DNA, separating two annealed nucleic acid strands.
Reference gene: gene or group of genes normally involved in the regulation/ maintenance of basic cell processes. It is assumed that the expression of these genes is the same independently on the external stimuli.
Hydrogen bonds: is the attractive interaction of a hydrogen atom with an electronegative atom such as nitrogen, oxygen or fluorine that comes from another chemical group.
Hydrophobic interactions: propensity of nonpolar substances to aggregate inaqueous solution and exclude water molecules.
Hyphae/Pseudohypae: is a long, branching filamentous structure of some Candida species.
Immunocompromised: is a state in which the immune system of a person is not functioning properly.
Initial colonization: the spreading of an initial microorganism into a new habitat.
Lagging DNA strand: is the DNA strand that grows in the opposite direction to the movement of the growing fork, and is synthesized discontinuously.
Leading DNA strand: is the DNA strand that is being synthesized continuously moving towards the replication fork.
Long-term urinary catheterization: the catheter remains in situ for more than 28 days. Long-term catheterization might be used in

patients without mobility or with urinary incontinence and severe illness.

Luciferase (Lux): protein produced by fireflies that is responsible for their luminescence. Like GFP, Lux can be used as a reporter protein for gene expression and protein localization studies.

Matrix: extracellular material mainly produced by the microorganisms themselves and that is involving and protecting them from harsh environment.

Melting curve: is the thermal denaturation profile, in molecular biology DNA melting curve allows the detection of mutations and other changes in the genetic code.

Microarrays: technique for analysis of the complete cellular transcriptome incorporating hybridization of fluorescently labelled cDNAs of the transcriptome to genomic DNA chemically cross-linked to a glass slide.

Microflora: the community of microorganisms, including bacteria, fungi, protozoans and algae that live in or on another living organism or in a particular habitat.

mRNA half-life: amount of time needed to loss 50% of the mRNA functionality.

mRNA transcripts: all the mRNAs produced by the cell in response to a stimuli.

mRNA: or messenger RNA are molecules that are a complementary "blue print" of the genetic information encoded in the cell genome.

Multidrug-resistance: tolerance of a microorganism to at least three classes or subclasses of antimicrobial agents that were originally effective for treatment of infections caused by it.

Multiplex cDNA libraries: cDNA libraries prepared for RNA sequencing using primers tagged with a barcode.

Non-specific priming: when primers bind to unspecific regions of DNA.

Northern blot: technique used to detect the steady-state levels of an RNA transcript at any given time within the cell. Can be thought of as a snapshot of an RNA transcript amounts within a total RNA prep.

Nosocomial infections: an infection that is typically acquired in a hospital or a healthcare service unit.

Nucleotides: or nucleic acids, are the building blocks of DNA and RNA. In DNA they are Adenine, Thymine, Guanine and Cytosine, and in RNA they are the same with the exception of Thymine which is replaced by Uracil.

Operon: functioning unit of genomic DNA containing a cluster of genes under the control of a single regulatory signal or promoter.

Opportunistic pathogen: a microorganism that under normal conditions is harmless to the human body but that can become a health threat if the immune system fails.

Organellar genomes: these are genomes contained inside cell organelles such as mitochondria or chloroplasts.

Origin of replication: is a particular sequence in the genome that marks where the DNA is going to begin to be replicated.

Osteomyelitis: is an infection/inflammation of the bone, caused by a microorganism.

Pathogenic bacteria: bacteria that is able to cause diseases to humans.

Persistent or chronic infections: infections characterized by the continuous presence of the infectious organism, often latent (with little or no symptoms) with occasional relapses of active infection.

Phenotype: collection of characteristics that are observable by the investigator. All genes are not expressed at the same time, and the environment profoundly influences phenotypic expression.

Planktonic: bacteria that are suspended or growing in a fluid environment.

Plasmid: DNA molecule that can replicate independently of the chromosomal DNA and encodes for non-essential genetic information, like antibiotic resistance for example.

Plastids: organelles found in photosynthetic organisms such as plants and algae and are sites for the manufacture and storage of chemical compounds needed by the cell.

Pleiotropic regulator: regulator that has effects on several different characters (one single gene is responsible for the regulation of several phenotypic characteristics).

Polymerase chain reaction: or PCR, is a technique used in molecular biology where the enzyme DNA polymerase is used in order to amplify a single or a few copies of DNA by several orders of magnitude generating several millions of copies of the desired sequence.

Polypeptides: short polymers of amino acids (protein's basic unit) held together by peptide bonds.

Primer dimer: process where primers hybridize with each other.

Probiotic: live microrganismos which, when ingested in adequate amounts, confer a health benefit to the host.

Prokaryotes: group of organisms that lack an enclosed cell nucleus, such as bacteria.

Proteome: complete set of proteins. It should correspond to the mRNAs in the transcriptome, although there can be differences of detail reflecting changes in the relative abundance or stabilities of mRNAs and proteins. It can be used to refer to the set of proteins coded by the whole genome or produced in any particular cell or tissue.

Psychrotrophic: that is able to survive or even develop in a cold environment.

Public health: the field of study dealing with the protection and improvement of community health by organized community effort and including preventive medicine, sanitary and social science.

Pure culture: a population of cells that have derived from the same single cell.

Regulons: group of genes that are regulated under the same condition or by the same regulatory protein.

Relapsing infections: reappearance of symptoms and signs of a disease after treatment.

Replication: or DNA replication, the process of copying double-stranded DNA.

RNA: ribonucleic acid or RNA is the prime molecule for the transfer of the information contained in the genome of organisms into proteins.

RNA polymerase: acts similar to DNA polymerase but helping in the polymerization of RNA strands.

RNA primer: strand of nucleic acid (RNA) that serves as a starting template for DNA replication.

RNA sequencing: technique for analysis of the complete cellular transcriptome. This method includes constructing a cDNA library of a complete transcriptome and the sequencing of this cDNA library using next-generation sequencing technology.

RNase: enzyme that degrades all species of RNA.

rRNA: or ribossomal RNA, is the RNA component of ribosomes where protein synthesis occurs.

Saprophytism: a relation in which an organism feeds on the dead remains or products of other organisms.

Semi-conservative replication: process of DNA replication whereby the two strands that make up the DNA separate and each of them (parent strains) act as a template for the formation of new strands (daughter strands) which are synthesised de novo.

Septicemia: systemic illness due to invasion of the bloodstream by bacteria coming from a local infection.

Short-term urinary catheterization: the catheter is used for a short period of time, until the patient regains the ability to urinate on his or her own.

Sigma factor: a molecule that enables the initiation of transcription, in prokaryotes, by enabling the specific binding of RNA polymerase to gene promoters.

Standard curve: serial dilutions of a molecule with known quantity that can allow us to: 1) determine the quantity of an unknown sample by extrapolation or, 2) the amplification efficiency of the qPCR reaction.

Sterilization: process by which microorganisms are killed or inactivated.

Surfactants: compounds that lower the surface tension of a liquid.

Syndrome: a group of symptoms that collectively indicate or characterize a disease, a psychological disorder or another abnormal condition.

Tilling arrays: subtype of microarray chips that differ from traditional microarrays in the nature of the probes (tilling arrays probe intensively for sequences which are known to exist in a contiguous region).

Topoisomerase: enzyme that unwinds and winds DNA in order for DNA to control protein synthesis and to facilitate DNA replication.

Transcription factor: sometimes called a sequence-specific DNA-binding factor) is a protein that binds to specific DNA sequences, thereby controlling the rate of transcription of genetic information from DNA to messenger RNA.

Transcription terminator: section of genetic material that signals the end of transcription of a certain gene.

Transcription: process of creating a complementary RNA copy of a sequence of RNA.

Transcriptome: strictly, is defined as all RNAs (coding and also non-coding), however, generally, refers to all mRNAs of a single or group of cells that are responding to environmental or internal stimuli.

Transduction: process where genetic material can be transferred from one bacterium to another.

Translation: process through which mRNA produced during transcription is decoded by the ribosome to produce a specific amino acid chain, or polypeptide, which will later fold into an active protein.

Transposable elements: or transposons are sequences of DNA that can move or "transpose" themselves in a new position within the cell genome.

tRNA: or transfer RNA, is an adaptor molecule of RNA that is important in the protein synthesis in ribosomes.

Upregulation: process by which the cells increase the quantity of a cellular component.

Urinary catheterization: insertion of a catheter into a patient's bladder via the urethra for temporary or permanent drainage of urine.

Viral capsid: is the protein capsule of a virus.

Viral genomes: viruses contain a variety of genetic material. The viral genome can be composed of DNA or RNA which in turn can be single-stranded or double-stranded.

Virulence factor: molecules expressed or secreted by pathogens which enable them to colonize, evade or inhibit the immune system, obtain nutrients or enter inside cells of the host.

About the co-authors

Ana Henriques has a BSc in Applied Microbiology from the University of Glamorgan (UK) and a PhD from the University of Wales (UK). She is currently a post-doc researcher at the Center for Biological Engineering of the University of Minho, Portugal. Her research interests include mechanisms of antimicrobial resistance and alternatives antimicrobial agents. She currently works on the study of the transfer of resistance genes within mixed species biofilms relevant in the food industry.

Ana M. Sousa has an MSc in Biomedical Engineering from the University of Minho. Currently, she is a PhD student, at the same university, working in Pseudomonas aeruginosa adaptation and evolution to cystic fibrosis lungs during infection development.

Anabela Borges has a BSc in Biology and an MSc in Clinical Biology from University of Trás-os-Montes e Alto Douro and a PhD in Chemical and Biological Sciences in the same university. Currently, she is a postdoctoral researcher in the Department of Chemical Engineering of the Faculty of Engineering of the University of Porto. Her main research activities are related to the problematic of bacterial resistance to conventional antibiotics and biofilm formation.

Andreia Azevedo has a BSc in Genetics and Biotechnology and an MSc in Comparative and Technological Molecular Genetics from the University of Trás-os-Montes e Alto Douro. Currently, she is a PhD student in the Department of Chemical Engineering, Faculty of Engineering held at the University of Porto. Her main interests are to understand the role of pathogenic and atypical bacteria in mixed species biofilms associated to urinary tract infections.

Ângela França has a BSc in Biology and an MSc in Clinical Biology, both from the University of Trás-os-Montes and Alto Douro, and a PhD in Biomedical Engineering from the University of Minho. Currently, she is a post-doc researcher at the Centre of Biological Engineering, University of Minho. Her main interest is the characterization of the molecular interactions between S. epidermidis biofilms and the host immune system.

Carina Almeida has a BSc in Applied Biology and a PhD in Biomedical Engineering, both at the University of Minho. Currently she is a postdoctoral researcher at the Center for Biological Engineering, University of Minho and the Laboratory for Process, Environment, Biotechnology and Energy Engineering, University of

Porto. She works with molecular techniques applied to microbial identification and with the characterization of multispecies biofilms.

Daniela Machado has a BSc in Public Health Technology, from the Superior School of Porto. Currently, she is a PhD student at University of Minho. Her main research interests are focused on the role of G. vaginalis in bacterial vaginosis and its prevalence in Portuguese pregnant women.

Diana Alves has an MSc in Biomedical Engineering from the University of Minho and she is currently a PhD student at the Center for Biological Engineering, at University of Minho. Her project aims to impart biomaterial surfaces with antibacterial properties using antimicrobial peptides and enzymes.

Diana Rodrigues has a BSc in Applied Biology and a PhD in Chemical and Biological Engineering, both from University of Minho. She is currently a post-doc researcher at the Centre for Biological Engineering, University of Minho. Her main research interests concern biofilms of pathogenic bacteria, being currently focused on the study of magnetic hyperthermia as an alternative method to control foodborne pathogens.

Eduardo Costa has a BSc and an MSc in Microbiology, both from School of Biotechnology of the Portuguese Catholic University. Currently, he is a PhD student at Center for Biotechnology and Fine Chemistry, at the same university. His main interests are to better understand the bioactive properties of a particular group of polysaccharides, chitosan and its derivatives, and its biological potential against pathogenic bacteria and their biofilms.

Fátima Lopes has a BSc in Biochemistry from the University of Lisbon and a PhD in Biochemistry from the Instituto de Tecnologia Química e Biológica (ITQB), New University of Lisbon. After a Post-doc in Microbiology, she got a position as Associate Researcher at ITQB. She has dedicated her work to study Enterococcus faecalis antibiotic resistances and pathogenesis processes in order to improve knowledge that will help finding new therapeutic and prophylactic actions to decrease enterococcal nosocomial infections.

Joana Castro has an MSc in Biomedical Engineering from University of Minho. She is currently a PhD student at the University of Porto. Her main research interests are to differentiate commensal and pathogenic G. vaginalis and their impact in bacterial vaginosis.

Joana Malheiro has a BSc in Biochemistry and an MSc in Molecular Genetics, both from University of Minho. Her main research activities focus on the evaluation of the potential of phytochemicals in disinfection, alone or in combination with commonly used biocides, against both planktonic and biofilm bacteria.

Luís F. Melo has a BSc in Chemical Engineering, from the University of Porto and a PhD Engineering Science-Transport Phenomena, from the University of Minho. He is a full professor in Chemical Engineering, at the University of Porto. His research interests include mainly the fields of Biofilm Science and Engineering, Heat Exchanger Fouling, Biological Wastewater Treatment and Drinking Water Systems.

Manuel Simões has a BSc in Biological Engineering and PhD in Chemical and Biological Engineering both from University of Minho. He is an Assistant Professor and member of the Laboratory for Process Engineering, Environment, Biotechnology and Energy in the Department of Chemical Engineering of the Faculty of Engineering of the University of Porto. His main research interests are focused on the mechanisms of biofilm formation and their control with antimicrobial agents, particularly with the use of new antimicrobial molecules.

Manuela Pintado has a BSc in Pharmacy, from the University of Porto and a PhD in Biotechnology from the Portuguese Catholic University, Porto. She is an Assistant Professor of the College of Biotechnology of the Portuguese Catholic University and the Coordinator of Department of Scientific Services for Industry. Her research interests include Biotechnology and Biochemistry - carry out research in the development and compositional characterization and validation of bioactivity of functional ingredients in particular those with antimicrobial and anti-biofilm activity.

Maria E. Rodrigues has an MSc and a PhD in Biomedical Engineering both at University of Minho. She currently is a post-doctoral researcher at the Centre for Biological Engineering of University of Minho, in the areas of fungal biofilms, microbiology and animal cell culture.

Maria O. Pereira has a BSc in Biological Engineering and a PhD in Chemical and Biological Engineering, both from the University of Minho. She is an Assistant Professor at the Department of Biological Engineering, University of Minho, Portugal. Her main research interests include the characterization of the molecular aspects underlying the pathogenesis of biofilm-associated microorganisms and the study of innovative approaches to successfully minimize the virulence aspects and resistance phenomena of pathogenic bacteria.

Mariana Henriques has a BSc and a PhD in Biological Engineering, both from the University of Minho. She is an Assistant Professor, at the Department of Biological Engineering, University of Minho. Her main research interests are focused in Candida virulence factors and antimicrobial surfaces.

Nuno F. Azevedo has a BSc in Biological Engineering and a PhD in Chemical and Microbial Technology, both at the University of Minho. After holding a Post-doctoral position at the University of Minho and the University of Southampton, he is currently a Faculty Researcher at the University of Porto. His research interests are focused on the development of fluorescence in situ hybridization methods for the detection of pathogens in clinical, environmental and food samples and is also employing these methods to discriminate different populations in multispecies biofilms.

Paula Jorge has a BSc in Biochemistry and an MSc in Bioengineering, both from the University of Minho. She is currently a PhD student in the same university. Her main interests are the use of antimicrobial peptides for the treatment of biofilm associated infections but also the application of bioinformatics tools to aid in the antimicrobial peptide area of research.

Sara Silva has a BSc in Biosciences and an MSc in Microbiology, both from School of Biotechnology of the Portuguese Catholic University. Currently, she is a PhD student at Center for Biotechnology and Fine Chemistry, at the same university. Her main interests are the study of the biological potential of natural compounds and their potential antimicrobial activity against several pathogenic microorganisms.

Sónia Silva has a BSc in Biology from Porto University, MSc in Molecular Genetics from Minho University and a PhD in Biomedical Engineering from the University of Minho. Currently she is an invited assistant researcher the Centre for Biological Engineering, University of Minho. She works with Candida infections and its associated virulent factors.

Susana Lopes has an MSc in Biological Engineering and a PhD in Chemical and Biological Engineering, both from the University of Minho. She is currently a post-doc researcher at the Center for Biological Engineering. Her main interests are the study of polymicrobial interactions within respiratory diseases (cystic fibrosis, ventilator-associated pneumonia).

Virginia Carvalhais has a BSc in Clinical Analysis and Public Health from the University Cooperativa de Ensino Superior Politécnico e Universitário and an MSc in in Molecular Genetics from the University of Minho. She is a PhD student at the University of Porto. Her main research interest is to study dormancy within Staphylococcus epidermidis biofilms.

CENTRE OF
BIOLOGICAL
ENGINEERING